ARAMIS or

THE LOVE OF TECHNOLOGY

ARAMIS or THE LOVE OF TECHNOLOGY

Bruno Latour

TRANSLATED BY CATHERINE PORTER

HARVARD UNIVERSITY PRESS 1996
Cambridge, Massachusetts, & London, England

Printed in the United States of America

This book was originally published as *Aramis, ou l'amour des techniques,* by Editions La Découverte, copyright © 1993 by Editions La Découverte, Paris.

Publication of this book has been aided by a grant from the French Ministry of Culture.

Library of Congress Cataloging-in-Publication Data

Latour, Bruno.
[Aramis. English]
Aramis, or the love of technology / Bruno Latour : translated by Catherine Porter.
p. cm.
ISBN 0–674–04322–7 (alk. paper)
ISBN 0–674–04323–5 (paperback : alk. paper)
1. Local transit—France—Paris Metropolitan Area. 2. Personal rapid transit—France—Paris Metropolitan Area. I. Title.
HE4769.P3L3813 1996
388.4′0944′361—dc20 95–34920

CONTENTS

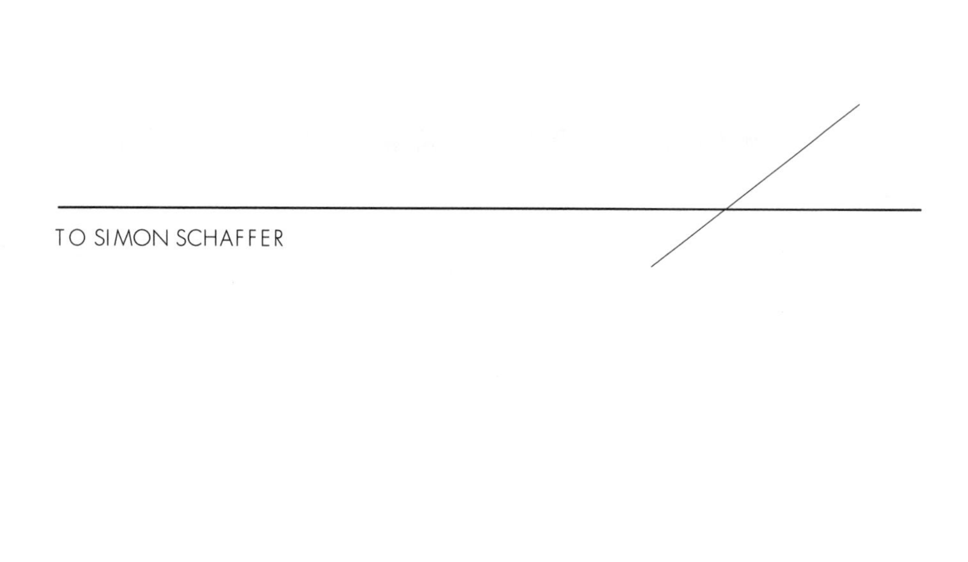

TO SIMON SCHAFFER

PREFACE

Can we unravel the tortuous history of a state-of-the-art technology from beginning to end, as a lesson to the engineers, decisionmakers, and users whose daily lives, for better or for worse, depend on such technology? Can we make the human sciences capable of comprehending the machines they view as inhuman, and thus reconcile the educated public with bodies it deems foreign to the social realm? Finally, can we turn a technological object into the central character of a narrative, restoring to literature the vast territories it should never have given up—namely, science and technology?

Three questions, a single case study in scientifiction.

Samuel Butler tells the story of a stranger passing through the land of Erewhon who is thrown into prison because he owns a watch. Outraged at the verdict, he gradually discovers that draconian measures forbid the introduction of machinery. According to the inhabitants of Erewhon, a cataclysmic process of Darwinian evolution might allow a simple timepiece to give birth to monsters that would rule over humans. The inhabitants are not technologically backward; but they have voluntarily destroyed all advanced machines and have kept none but the simplest tools, the only ones compatible with the purity of their mores.

Butler's Nowhere world is not a utopia. It is our own intellectual universe, from which we have in effect eradicated all technology. In this universe, people who are interested in the souls of machines are severely punished by being isolated in their own separate world, the world of engineers, technicians, and technocrats.

By publishing this book, I would like to try to bring that isolation to an end.

I have sought to offer humanists a detailed analysis of a technology sufficiently magnificent and spiritual to convince them that the machines by which they are surrounded are cultural objects worthy of their attention and respect. They'll find that if they add interpretation of machines to interpretation of texts, their culture will not fall to pieces; instead, it will take on added density. I have sought to show technicians that they cannot even conceive of a technological object without taking into account the mass of human beings with all their passions and politics and pitiful calculations, and that by becoming good sociologists and good humanists they can become better engineers and better-informed decisionmakers. An object that is merely technological is a utopia, as remote as the world of Erewhon. Finally, I have sought to show researchers in the social sciences that sociology is not the science of human beings alone—that it can welcome crowds of nonhumans with open arms, just as it welcomed the working masses in the nineteenth century. Our collective is woven together out of speaking subjects, perhaps, but subjects to which poor objects, our inferior brothers, are attached at all points. By opening up to include objects, the social bond would become less mysterious.

What genre could I choose to bring about this fusion of two so clearly separated universes, that of culture and that of technology, as well as the fusion of three entirely distinct literary genres—the novel, the bureaucratic dossier, and sociological commentary? Science fiction is inadequate, since such writing usually draws upon technology for setting rather than plot. Even fiction is superfluous, for the engineers who dream up unheard-of systems always go further, as we shall see, than the best-woven plots. Realism would be misleading, for it would construct plausible settings for its narratives on the basis of specific states of science and technology, whereas what I want to show is how those states are generated. Everything in this book is true, but nothing in it will seem plausible, for the science and technology it relies upon remain controversial, open-ended. A journalistic approach might have sufficed, but journalism itself is split by the great divide, the one I'm seeking to eliminate, between popularizing technology and denouncing its politics. Adopting the discourse of the human sciences as a master discourse was not an option, clearly, for it would scarcely be fitting to call the hard sciences into question only in order to start taking the soft ones as dogma.

Was I obliged to leave reality behind in order to inject a bit of emotion and poetry into austere subjects? On the contrary, I wanted to come close enough to reality so that scientific worlds could become once again what they had been: possible worlds in conflict that move and shape one another. Did I have to take certain liberties with reality? None whatsoever. But I had to restore freedom to all the realities involved before any one of them could succeed in unifying the others. The hybrid genre I have devised for a hybrid task is what I call *scientifiction*.

For such a work, I needed a topic worthy of the task. Thanks to the Régie Autonome des Transports Parisiens (RATP), I was able to learn the story of the automated train system known as Aramis. Aramis was not only technologically superb but also politically impeccable. There was no "Aramis affair," no scandal in the newspapers. Better still, during the same period the very same companies, the same engineers and administrators, succeeded in developing the VAL automated subway systems whose background forms a perfect counterweight to the complex history of Aramis. Even though I had not gone looking for it at the outset, the principle of symmetry hit home: How can people be condemned for failing when those very same people are succeeding elsewhere?

I could have done nothing without the openness and sophistication, new to me, of the world of guided transportation (that is, transportation that functions on rails). The few engineers and decisionmakers in this field, who have been renewing the framework of French urban life through spectacular innovations in public transportation over the last twenty years, were nevertheless willing to cooperate in the autopsy of a failure. It is owing to their openmindedness, with special thanks to the RATP, the Institut National de Recherche sur les Transports (INRETS), and Matra Transport, that Aramis can be presented to us all as an exemplary meditation on the difficulties of innovation. So Aramis will not have died in vain.

This book, despite its strange experimental style, draws more heavily than the footnotes might suggest on the collective work of the new sociologists of technology. Particularly relevant has been the work of Madeleine Akrich, Wiebe Bijker, Geoffrey Bowker, Alberto Cambrosio, Michel Callon, John Law, and Donald MacKenzie. Unfortunately, the book was published too soon for me to use the treasure trove of narrative resources developed by Richard Powers, the master of scientifiction and author of *Galatea 2.2,* whose Helen is Aramis' unexpected cousin.

Here is one more cue for readers:

In this book, a young engineer is describing his research project and his sociotechnological initiation. His professor offers a running commentary. The (invisible) author adds verbatim accounts of real-life interviews along with genuine documents, gathered in a field study carried out from December 1987 to January 1989. Mysterious voices also chime in and, drawing from time to time on the privileges of prosopopoeia, allow Aramis to speak. These discursive modes have to be kept separate if the scientifiction is to be maintained; they are distinguished by typography. The text composed in this way offers as a whole, I hope, both a little more and a little less than a story.

"IT'S TRULY A NOVEL, THAT STORY
ABOUT ARAMIS . . ."
"NO, IT'S A NOVEL THAT'S TRUE, A
REPORT, A NOVEL, A NOVEL-REPORT."
"WHAT, A FAKE LOVE STORY?"
"NO, A REAL TECHNOLOGY STORY."
"**NONSENSE!** LOVE IN TECHNOLOGY?!"

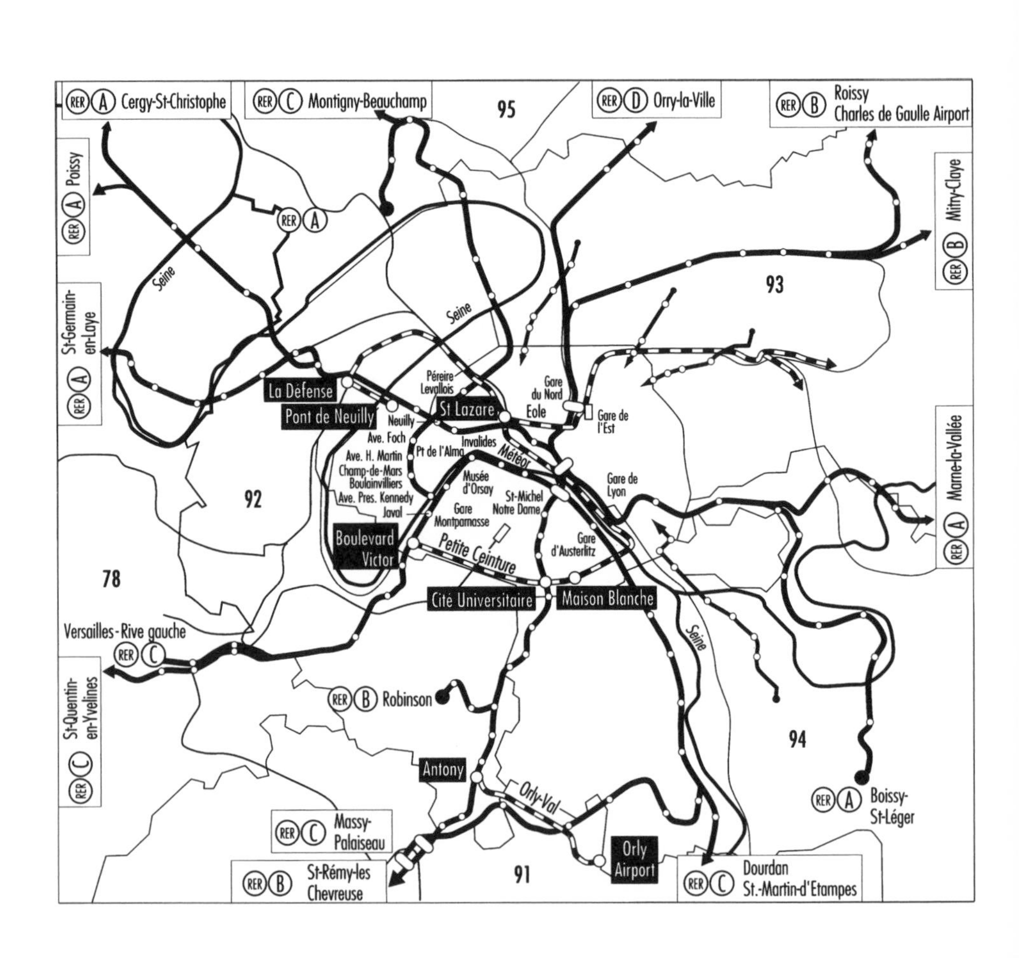

RER A Cergy-St-Christophe
RER C Montigny-Beauchamp
95
RER D Orry-la-Ville
RER B Roissy Charles de Gaulle Airport
RER A Poissy
RER A
RER B Mitry-Claye
93
RER A St-Germain-en-Laye
Seine
La Défense
Pont de Neuilly
Péreire-Levallois
St Lazare
Gare du Nord
Eole
Gare de l'Est
Neuilly
Ave. Foch
Ave. H. Martin
Champ-de-Mars
Boulainvilliers
Ave. Pres. Kennedy
Javal
Pt de l'Alma
Invalides
Météor
Musée d'Orsay
St-Michel Notre Dame
Gare de Lyon
Gare Montparnasse
Gare d'Austerlitz
92
RER A Marne-la-Vallée
Boulevard Victor
Petite Ceinture
78
Cité Universitaire
Maison Blanche
Versailles-Rive gauche
RER C
RER C St-Quentin-en-Yvelines
RER B Robinson
94
Antony
Orly-Val
RER A Boissy-St-Léger
RER C Massy-Palaiseau
Orly Airport
RER B St-Rémy-les Chevreuse
91
RER C Dourdan St.-Martin-d'Etampes

PROLOGUE: WHO KILLED ARAMIS?

The first thing I saw when I went into Norbert H.'s office was the new RATP poster on the wall [see Photo 1]:

[DOCUMENT: TEXT OF THE RATP'S ADVERTISEMENT LAUNCHING THE R-312 BUS]

Darwin was right!

RATP means the evolution and adaptation of buses in an urban environment.

In 1859 Darwin proposed his theory of evolution, maintaining that the struggle for life and natural selection should be seen as the basic mechanisms of evolution.

The latest product of this evolution is the R-312 bus, which is about to begin service on Line 38. For the occasion, today's buses and their predecessors will join in a big parade in honor of the R-312.

The theory of evolution has its advantages. Thanks to Darwin, you can ride our buses around the Luxembourg Garden for free on Wednesday, June 1.

"Chausson begat Renault, Renault begat Schneider, Schneider begat the R-312 . . . Darwin's theory has its downside," said my future mentor solemnly when he saw me reading the poster. "There are people who want to study the transformation of technological objects without worrying about the engineers, institutions, economies, or populations involved in their development. The theory of evolution can take such people for a ride! If you leave your engineering school to come study innovation, my friend, you'll have to drop all that third-rate biology. This may disappoint you, but—unless I'm completely incompetent in such matters—a bus does not have sex organs. Never mind the poster: the R-312 doesn't descend from the Chausson APU 53 the way humans descend from apes. You can climb aboard a bus, but you can't climb back to the Schneider H that was all over Paris in 1916. Frankenstein's monster with his big dick and his lopsided face? Such things exist only in novels. You'd have quite a crowd of people parading around the Luxembourg Garden if you really wanted to honor all of the new bus's progenitors."

I hadn't yet done any in-depth studies of technological projects. I'd just emerged from a telecommunications school where I'd taken only physics and math; I'd never seen a motor, or a chip, or even the inside of a telephone. That's why I wanted to spend a year at the Ecole des Mines, in sociology. There at least, or so I'd been told, ambitious young people could learn the engineering trade and study real projects in the field. I didn't find it at all reassuring to be abandoning the peace and quiet of technological certainties only to apprentice myself to a laboratory Sherlock who'd just been entrusted by the RATP with the investigation of a recent murder: "Who killed Aramis?" I'd read *The Three Musketeers,* but I didn't know Aramis and wasn't aware he was dead. In the beginning, I really thought I'd landed in a whodunnit, especially since Norbert, the inspector to whom I'd been assigned, was a fellow at least forty years old with a Columbo-style raincoat.

"Here's the beast," my professor said [see Photos 11–14]. "It's a new transportation system, apparently a brilliant design. A combination of private cars and public transportation. The ideal, you might say. In any case, it's not like the R-312; there wasn't any parade in Aramis' honor, and there certainly weren't any Darwinian posters. Just a slightly sad farewell party on the boulevard Victor, at the site of the Center for Technological Experimentation (CET) three weeks ago, in early December 1987. A promising, seductive, dazzling line of technology has been buried without fanfare. The site will be an empty lot for a while, until it's developed as part of the renovation of the quai de Javel. You should have seen how mournful the engineers were. According to what they told me, the project was really admirable. They'll never have another chance to build, from the ground up, an entirely automatic and entirely revolutionary system of guided transportation—a system running on rails. But Aramis fell out of favor. 'They dropped us'—that's what the engineers say. 'They' who? The Nature of Things? Technological Evolution? The Parisian Jungle? That's what we've been asked to find out, my friend, because we don't belong to the transportation world. Some people claim that Aramis wouldn't have kept its promises. But others, apparently, say that it was the State that didn't keep its promises. It's up to us to sort all this out, and we can't rely on Darwin or on sexual metaphors. And it won't be easy."

Personally, I didn't see the problem. I replied confidently that all we had to do was take a close look to see whether the project was technologically feasible and economically viable.

"That's all?" asked my mentor.

"What? Oh, no, of course not; it also has to be socially acceptable."

Since my professor was a sociologist, I thought I was on the right track. But he grinned sardonically and showed me his first interview notes.

[INTERVIEW EXCERPTS]

"It doesn't make any sense. Six months ago, everybody thought it was the eighth wonder of the world. Then all of a sudden everything fell apart. Nobody supported it any longer. It happened so fast that no one can figure it out. The head of the company can't figure it out either. Can you do something? Say something? . . ."

"It had been going on for twenty years; the time had come to call it quits. It'll be a fine case for you muckrakers from the Ecole des Mines. Why did they keep that monstrosity going so long on intravenous feedings, until somebody finally had the balls to yank out the tubes? . . ."

"It's typically French. You have a system that's supposedly brilliant, but nobody wants it. It's a white elephant. You go on and on indefinitely. The scientists have a high old time . . ."

"That's France for you. You get a good thing going, for export; it's at the cutting edge technologically; people pour money into it for fifteen years; it revolutionizes public transportation. And then what happens? The Right comes to power and everything comes to a screeching halt, with no warning, just when there's finally going to be a payoff. It would really help if you could do something about it. Why did they drop a promising project like this after supporting it for so long? . . ."

"The industrial developer let it go. They got their studies done at our expense; then it was 'Thank you' and 'Goodbye' . . ."

"The operating agency couldn't accept an innovation that was the least bit radical. Corporate culture is the problem. Resistance to change. Rejection of a transplant . . ."

"The public authorities are losing interest in public transportation. It's another ploy by the Finance Ministry, business as usual . . ."

"It's an economic problem. It was beautiful, but it cost too much. So there was no choice . . ."

"It's old-fashioned. It's backward-looking. It's the sixties. In 1987 it's no good, it won't fly . . ."

"In ten years—no, five—it'll be back, take my word for it. It'll have a new name; but the same needs create the same technologies. And then people will really kick themselves for abandoning it just when everybody would have wanted it . . ."

"But what's the real answer?" I asked with a naïveté that I regretted at once.

"If there were one, they wouldn't pay us to find it, chum. In fact, they don't know what killed Aramis. They really don't know. Obviously, if by 'real answer' you mean the official version—then, yes, such versions exist. Here's one."

[DOCUMENT: EXCERPTS FROM AN ARTICLE PUBLISHED IN *ENTRE LES LIGNES*, THE RATP HOUSE ORGAN, JANUARY 1988]

Four questions for M. Maire, head of research and development.

Do transportation systems like Aramis really fill a niche, from the user's point of view?

The idea of little automated cabs that provide service on demand is seductive *a priori*, but hard to bring off economically. Furthermore, the creation of a new mode of transportation is a tricky business in a city where billions of francs have been invested in the infrastructures of other transportation systems that do the job perfectly well. In new cities or in cities that don't have their own "on-site" transportation, a system like Aramis can offer an interesting solution. The project designed for the city of Montpellier would be a good example, except that there, too, implementation had to be postponed for financial reasons.

People talk about the failure of the Aramis project. But can't it be seen as a success, given that the experimental card was played and appropriate conclusions were drawn?

It's not a failure; on the contrary, it's a technological success. The CET has demonstrated that the Aramis principles were valid and that the system could work. We did play the card of experimentation, there's no doubt about it. But the evolution of needs and financial resources doesn't allow for the implementation of such a system to be included among the current priorities for mass transportation in Paris. Why would you want us to keep on trying to perfect a transportation system that we see no real use for in the short run, or even in the medium run?

The Aramis CET was the first phase of a project that was intended to serve the southern part of the Petite Ceinture in Paris. The problem of providing this service still hasn't been resolved. Aren't there some risks involved in coupling a research project like this with a project for upgrading the transportation network?

The important thing now is to protect the existing track

system of the Petite Ceinture so as to avoid mortgaging the construction of a future public transportation line. Anyway, some market studies will have to be redone, perhaps with an eye toward liaison with an automated mini-metro. As for the notion of risk, I don't agree. If we don't try things, we'll never accomplish anything new. Generally speaking, it stimulates research if you have concrete objectives. It also makes it easier to mobilize decisionmakers around a project—even if there's some risk in doing so!

Aramis comes across as a technological gamble. Do the studies that have been carried out give Matra Transport and the RATP a head start in the realm of automated urban transportation?

Even if the Aramis project wasn't initially intended to be a melting pot for new urban transportation technologies, it ended up playing that role. There will be a lot of spillover. Besides, research has shown how important it was to take a global approach in thinking about the transportation of tomorrow. The key to success is as much in the overall vision of the system as in mastery of the various technological components.

I wasn't used to making subtle distinctions between technical feasibility and "official versions" of what is feasible or not. I'd been trained as an engineer. I didn't really see how we were going to go about finding the key to the enigma.

"By going to see everybody who's being criticized and blamed. Nothing could be simpler."

My boss had his own peculiar way of going about these things. In the evening, after the interviews, he would organize "meetings and confrontations" (as he called them) in his file-cluttered office. What he actually did was arrange our interview transcripts in little bundles.

"That's the big difference between sociology and justice. They don't come to us; we go to them. They answer only if they feel like it, and they say only what they want to say."

"You see," he went on during one of these daily "confrontations," "there have been hardly any questions about the proximate causes of Aramis' death. It all happened in three months."

[INTERVIEW EXCERPTS]

The scene is the RATP premises on the boulevard Victor, in December 1987, three hundred yards from the workshop where the five Aramis prototype cars sit motionless. The project engineers are talking heatedly:

"While a meeting was under way in February 1987, M. Etienne *[of Matra]* secretly distributed a 'provisional verbal note' (it was in writing, all the same) saying 'Stop everything.' Frankly, we didn't understand what was going on." [no. 2]*

M. Girard, in a temporary office downtown:

"The end didn't surprise me. The Finance Ministry was all it took . . . We had a colossus with feet of clay. Its whole support structure had disappeared in the meantime . . .

"It hardly matters who was responsible for piling on the last straw; that was just the proximate cause. In any event, the point is that *all it took was one last straw.* It doesn't matter who killed the project. As for the proximate cause, I don't know."

"But you know the remote cause?"

"Yes, of course. Actually, when I realized that Aramis had been called off, it didn't surprise me. For me, it was *built right into the nature of things.*" [no. 18]

M. Desclées, in an elegant suburban office of the Institut National de Recherche sur les Transports (INRETS):

"There's one thing I don't want to see glossed over in your study . . . There was a very important political change after 1986.† Soulas, the new RATP president, had been general inspector of finances, whereas Quin's experience

*The numbers refer to the original interviews. Certain protagonists were interviewed several times. Some interviews were conducted in a group setting. Certain data come from sessions devoted to summing up the investigation for the benefit of the client; these sessions are called "restitutions."

†The legislative election brought the Right to office for a two-year period of power sharing between President François Mitterand and Prime Minister Jacques Chirac.

was in marketing and public relations. The new president wanted to bring all superfluous research to a halt. After a few months I went to see him; he told me to 'cut all that out.' I said, 'When you've already spent 95 percent of your budget, maybe it's best to keep going.'

"Soulas got the endgame under way when he told Etienne that the construction of the line would not be included in the next Five-Year Plan; this was in late 1986 or early 1987.

So the RATP people may tell you that they 'don't understand what happened,' but the first blow came from within their own ranks." [no. 11]

M. Frèque, one of Matra's directors, speaking at Matra headquarters:

"By late 1986 I'd become convinced that it had to stop . . . Our conclusions were increasingly negative. Production costs were going up, with harmful results for us because the State's participation was constant whereas ours was variable.

"So as early as the twenty-seventh month we were going in a different direction from the protocol. The others were saying, 'Finish your product and do what you can with it. Later on, we'll see about building the line.'

"Read the protocol: by the twenty-seventh month we were supposed to be *in production!* In my first report, this was clearly spelled out; later on, they glossed over it.

"That was it, for me. I'd faith in this thing. We came to an agreement. The testing team worked on November 11, a legal holiday, and I'm very proud of that. When the ship is going down, you stay at your post until the last minute—that's something I believe in." [no. 6]

Matra headquarters again. M. Etienne, the president is speaking:

"What changed everything was the change of president *[of the RATP]*. He came in June; I met him in October 1986. He said, 'Give me time.' I took him to Lille on October 26 to see VAL. I remember I sent him a note. 'Here's what we think: we don't have any major applications; the system has to be simplified; the network isn't complicated enough to justify a complicated system.'

"He told me, 'There'll be practically nothing in the Tenth Plan for the RATP—not for Aramis at any rate, not for construction of the line.' I thanked him for his frankness. Now that I've gotten to know him, he's very straightforward. 'Nothing will happen for the next seven years.

". . . The RATP was accepting the cutbacks, very quietly. We were wondering how far they were going to allow this thing to go.

"But during that period I found out for certain that neither the DTT nor the finance minister intended to contribute a thing. Soulas was right.

"This is where my February 1987 note came from. Nothing was going to happen for seven years. 'Matra wants to pull out early'—that's what the RATP people were saying.

"What we were saying here was: 'Let's refocus Aramis, make it more efficient.' We wanted to renegotiate. 'When we start up again seven years from now, we'll at least have something to start with.'

"They sulked. 'Since you want to pull out, we'll shut the whole thing down.' That wasn't what we wanted. But the State and the public authorities didn't have any more money.

"With patience, Soulas got the reforms he was after. He got Aramis shut down; he was totally honest." [no. 21]

M. Maire, one of the directors in charge of the RATP's research and development, speaking at the agency's headquarters:

"Etienne showed VAL to Soulas, and Soulas saw his chance to ask: 'What about Aramis?' 'It won't get a cent.' 'I get the picture,' Etienne replied."

"So the final decision really did come from the RATP?"

M. Etienne: "No, no, not at all. Soulas was the mouthpiece for the Finance Ministry. For them, any innovation is a drain on resources. It certainly wasn't a question of Soulas' being won over by the RATP, or by our people, our engineers." [no. 22]

M. Soulas, president of the RATP, in his plush second-floor office overlooking the Seine:

"Aramis died all by itself, Professor H. I didn't intervene. I can say this quite freely, because I'm an interventionist president and I'd tell you if I'd stepped in. I didn't understand what was happening. It had been on track for fifteen years.

"It was a seductive idea, Aramis—really quite ingenious. It wasn't a line like a subway, but more like a bloodstream: it was supposed to irrigate, like veins and arteries. Obviously the idea doesn't make sense if the system becomes a linear circuit—that is, if it ceases to be a network.

"But this good idea never found a geographic footing. It was abstract. In its linear form, it tended to get transformed into a little metro; as a system, it became increasingly hybrid and complicated. Many people admired it. It became more and more technical, less and less comprehensible to the uninitiated, and a source of anxiety for the Finance Ministry. I watched it die. I didn't intervene in its death; I didn't have to."

"But you pushed a little, didn't you?"

"No, I didn't need to push. I found out one day that Aramis was being jettisoned. Matra had decided, or the RATP technicians . . . I'd be interested to know whose decision it was, actually. My feeling is that Matra made the fatal move. In any event, the top priority now is adding a line that will parallel Line A of the RER. Before, we could afford to experiment with Aramis; now we can't even manage a night on the town, as it were."

[Settling back more comfortably in his armchair.] "It's extraordinary that they've asked you to do this study! You know what it reminds me of? Oedipus' asking the soothsayer why the plague has come to Thebes! . . . You have the answer in the question itself. They wear blinders. Oh, there's not an ounce of ill will among the lot of them, but I've never seen such unpolitical people. 'How could it happen?' they must be wondering. 'How could we make something that works, and then it all goes belly-up?' It's touching, really—shows an extraordinary lack of awareness. Their own case intrigues them, because they like sociology . . . Aramis is such an intricate mess, incredibly intricate." [no. 19]

"You see, my friend, how precise and sophisticated our informants are," Norbert commented as he reorganized his notecards. "They talk about Oedipus and about proximate causes . . . They know everything. They're doing our sociology for us, and doing it better than we can; it's not worth the trouble to do more. You see? Our job is a cinch. We just follow the players. They all agree, in the end, about the death of Aramis. They blame each other, of course, but they speak with one voice: the proximate cause of death is of no interest—it's just a final blow, a last straw, a ripe fruit, a mere consequence. As M. Girard said so magnificently, 'It was built right into the nature of things.' There's no point in deciding who finally killed Aramis. It was a collective assassination. An abandonment, rather. It's useless to get bogged down concentrating on the final phase. What we have to do is see who built those 'things' in, and into what 'natures.' We're going to have to go back to the beginning of the project, to the remote causes. And remember, this business went on for seventeen years."

"There's one small problem," I said timidly. "I don't know a thing about transportation."

"Neither do I," replied my boss serenely. "That's why I was chosen. In a year, you can learn about any subject in the world. There's work ahead, but it will be good for your education. You're going to lose your innocence about the sexuality of technology, Mister Young Engineer. And I'm going to take advantage of the opportunity by writing a little commentary, a little sociology manual to make your work easier. You're to read it in addition to the books on this list; they're all in the school library."

He put on his old raincoat and disappeared in the drizzle, heading down the boulevard Saint-Michel.

Left to my own devices, I looked at the list. It included eighty-six titles, two-thirds of them in English. Tell an engineer to read books? It was quite a shock. As for the commentary, I was certainly going to need it, because there was a further complication: the laboratory where I was doing my internship used the word "sociology" in a way that absolutely no one else did.

1 AN EXCITING INNOVATION

"As it turns out," Norbert told me, "we're going to get a lot of help from a retrospective study done by the RATP. Here's a chronological chart that sums up the project's phases starting in 1970. Each phase is defined by its code name, by the money spent (in constant francs), and by its time frame. The horizontal axis shows annual expenditures. You can see they hesitated a lot. And the point at which they were spending the most money is the point at which everything ground to a halt, in 1987."

"It stopped before 1981, you'd have to say. And after Mitterand's election it started up again. Then, after Chirac's government came in, it fell off again . . ."

"That's right, my friend, elections do count in technology. You didn't suspect that?"

"Uh, well, yes," I responded prudently. "So we're beginning with the preliminary phase?"

"Yes, this one, right before Phase 0."

PROJECT CHRONOLOGY

1969: DATAR enlists Bardet's company Automatisme et Technique, for a study of various Personal Rapid Transit systems.

1970: Matra buys patents from Automatisme et Technique.

1973: Test site at Orly; three-vehicle train; demerging; merging; regrouping; four seats; off-line stations; passenger-selected destination; Transport Expo exhibit in Washington.

1974, February: Final report on Phase 0; creation of the Aramis development committee.

1974, May: Beginning of Phase 1; site analyses for the South Line; eleven sites studied; six seats; concept of point-to-point service abandoned; end of wholesale use of off-line stations. Giscard d'Estaing elected president; Bertin's aerotrain abandoned; tramway competition initiated.

1975: Variable-reluctance motor.

1976: Final report on Phase 1; Aramis simplified for economic reasons.

1977: Beginning of Phase 2A; Aramis simplified; ten seats; site analyses in Marne-la-Vallée; VAL marketed in Lille.

1978: Final report on Phase 2A; beginning of Phase 3A; test of the system's main components; testing grounds established; site analyses at La Défense and Saint-Denis, on the Petite Ceinture, and elsewhere.

1980: Final report on Phase 3A.

1981: Teams disbanded; no activity. Mitterand elected president; Fiterman named transportation minister; Quin becomes president of the RATP.

1982: Team reconstituted from VAL teams; site analyses in Dijon, Montpellier, Nice, and Toulon, and on the Petite Ceinture; Araval proposal, Aramis greatly simplified; initiation of SACEM project. Phase 3B: two-car units, twenty passengers; new test runs at Orly; initiation of the project for the World's Fair.

1983: Favorable final report on Phase 3B; VAL put into service; World's Fair project abandoned (June); site analyses in Montpellier for an Aramis using the VAL automation system.

1984, July: Protocol for construction of CET signed; Fiterman and Communist ministers leave the administration.

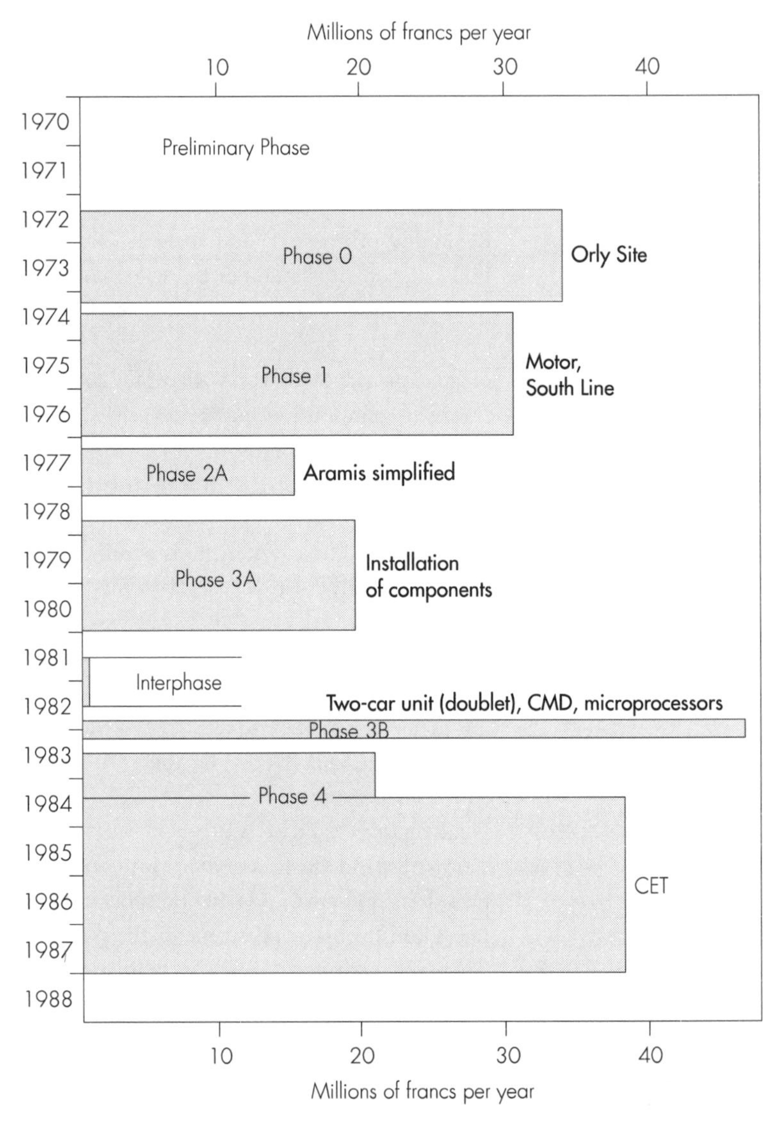

Figure 1. Total and annual expenses by phase, in 1992 francs. After Phase 3A, the amounts no longer include either the RATP's internal expenses or Matra's cost overruns. (Official contracts, January 1, 1988.)

1985: Scale model of two-car Aramis presented. 1986: First two-car unit delivered; two-tiered Aramis proposed; studies of potential ridership; Chirac named prime minister; Quin leaves the RATP.

1987: Termination of project announced; fifth two-car unit delivered; three weeks of contradictory test runs; project halted; postmortem study begun.

[INTERVIEW EXCERPTS]

M. Liévin, INRETS engineer:

"Aramis is the last of the PRT systems, you know."

"PRT?"

"It was fashionable at the time—Personal Rapid Transit, PRT. Everybody was excited about it."

"When was this?"

"Oh, around the sixties. The Kennedy era. Private cars were on the way out—that's what everyone was saying. But at the same time the advantages of cars had to be maintained; we couldn't keep moving in the direction of mass transportation." [no. 15]

M. Etienne, at Matra:

"In 1972 the Transport Expo took place in Washington. That exhibit was critical for PRTs. Everybody in the world came. There were Boeing systems, there were Bendix systems. People were beginning to recognize the potential of computers. It seemed logical to control vehicles from a central computer." [no. 21]

M. Cohen, speaking at Matra's Besançon office:

"All the major manufacturers plunged into PRTs: Boeing, Otis, we did the same thing at Matra. There were at least ten different systems. None of them worked. Aramis is the one that lasted the longest; it was the most credible, finally. We wouldn't do it the same way today; we wouldn't tell ourselves that, well, we know how to make planes and satellites, great, mass transportation must be a cinch. It's not true; it's not that easy. A train may well be more complicated than a satellite, technologically speaking." [no. 45]

Mr. Britten, an American private consultant in Paris:

"With PRTs, what doesn't work is the P. *P* means *people,* not *personal.* We knew from the beginning that that part didn't have a chance. In 1975—I have the report right here, you can look at it—we said that the only thing that made a difference was government support.* Either the project gets continuous government support, or else the whole family of Aramis-type PRTs collapses. It's that simple."

Summary of European PRT Project. (From ECOPLAN, *Innovational Guideway Systems and Technology in Europe* [Paris: Transport Research Group, January 1975].)

Project name	Year of origin	Government backing	Status in 1971–72	Present status
ARAMIS	1967	Yes	Advanced R&D	Active
CABINENTAXI	1969	Yes	Hardware development	Active
CABTRACK	1965	Yes	Advanced R&D	Abandoned
Coup	1969	None	Concept only	Abandoned
ELAN SIG	1970	None	System design	Under study
Heidt Automatischebahn	1971	None	Concept only	Abandoned
Schienentaxi	1975	None	Concept only	Abandoned
Spartaxi	1969	None	PRT site study	Set aside
TRP (Otis TTI)	1968	Some		Active
TRANSURBAN NONSTOP	1969	None	Preliminary design	Abandoned

"Here's an innovation with a niche that's easy to understand, for once," sighed my boss as he elbowed people aside so we could exit from the subway car. "If I take my car, I'm stuck for hours in traffic jams. If I walk, I breathe carbon dioxide and get lead poisoning. If I take my bike, I get knocked down. And if I take the subway, I get crushed by three hundred people. Here, for once, we have no problem understanding the engineers. They've come up with a system that allows us to be all by ourselves in a quiet little car, and at the same time we're

*See also Catherine G. Burke, *Innovation and Public Policy: The Case of Personal Rapid Transit* (Lexington, Mass.: Lexington Books, 1979).

in a mass transit network, with no worries and no traffic jams. That would be the ideal. I for one would welcome PRTs like the Messiah."

"Isn't it always that way?" I asked.

"You've got to be kidding! The last study I read was on inertial guidance systems for intercontinental missiles.* Those things are not greeted like the Messiah."

"You're right," I said, edging back into the cloud of smoke on the quai des Grands-Augustins.

[INTERVIEW EXCERPTS]

M. Parlat, speaking in the now-empty prefabricated building on the boulevard Victor that had been the Aramis project site:

"Aramis, the heart of Aramis, is nonmaterial coupling. That's the whole key. The cars don't touch each other physically. Their connection is simply calculated."

"Forgive my ignorance, but there's something I don't understand. Why don't they attach the cars together mechanically? *I mean, I don't know, with magnetic couplings, and then uncouple them automatically? They really don't know how to do it?"*

"No, it's out of the question. Eveything has been tried. We know how to do automatic couplings and uncouplings on stationary cars. We don't know how to couple and uncouple moving vehicles mechanically. Think about it: cars that are several meters long, going 30 kilometers an hour, coming up to a switch. Okay, this one takes the siding, that one keeps on going and links up with the car ahead. Mechanically, it's impossible. No, it can only be calculated, and even that isn't as simple as it sounds." [no. 3]

M. Liévin, speaking at INRETS:

"If you take trains made up of elements that can each go in a different direction, it's impossible to use mechanical couplings. Besides, there's a simple problem. Mechanical coupling transmits the force of all the other cars during braking and start-up. So each car has to be solid enough to stand up against the entire train. PRTs are lightweight vehicles—automobiles, nutshells. They can be light because they never touch each other, because they're connected

*D. MacKenzie, *Inventing Accuracy: A Historical Sociology of Nuclear Missile Guidance Systems* (Cambridge, Mass.: MIT Press, 1990).

electronically but not physically. That's the Aramis revolution: a major weight reduction. We've gone from railroads to automobiles, thanks to nonmaterial coupling." [no. 15]

M. Chalvon, at Alsthom:

"You never talk about mechanical uncoupling as a solution?"

"No, it doesn't exist. It's impossible. In any case, not if speed is a factor. It's not even an option. It just doesn't come up." [no. 46]

How to frame a technological investigation? By sticking to the framework and the limits indicated by the interviewees themselves.

They all say the same thing: "At the time, the world was dreaming of PRTs; mechanical uncoupling was impossible." For our informants, PRTs are no longer the invention of an isolated engineer, traceable through projects, contracts, and memoranda; rather, they're a collective dream. The technological impossibility of uncoupling is not a decision or the opinion of a handful of researchers. It's self-evident, obvious to everybody. Goes without saying. Doesn't generate the slightest controversy. It would take a Martian landing in the world of guided transportation to open up that question. Our interviewees no longer even manage to recall who might have come up with the dream of PRT. They can't tell you what institutions were behind its development. They can't come up with the names of the dozen or so engineers, journalists, middlemen, and public officials that would allow the investigator to replace the term "everybody" with a lobby, a school, a network. In 1988 the Sixties are remote. The origin of the project (1968, 1969) quickly gets lost in the mists of time, and like every narrative of origins it takes on the mythical characteristics of all Mists of Time: "Once upon a time; Everybody; No one can resist; Impossible." Of course, a historian of technology ought to work back toward that origin and replace it with groups, interests, intentions, events, opinions. She would go to America, to Germany, to Japan. She would visit the SNCF; she would work out the entire history of couplings and uncouplings. She would rummage through the archives. She would sketch the enormous fresco of guided transportation. She would reposition Aramis "in its historical framework"; she would determine its place in the entire history of guided-transportation systems. She would go further and further back in time. But then we would lose sight of Aramis, that particular event, that fiction seeking to come true. Since every study has to limit its scope, why not encompass it within the boundaries proposed by the interviewees themselves? None of them goes back further than 1965. For all of them, PRTs are beyond discussion: everyone wanted them; they had to be developed. There is no disagreement on this point. No engineer leaves open the possibility of mechanical uncoupling of cars. It's out of the question.

The investigator does not have to take the discussion any further. He will enjoy reading the historian, enjoy crossing the mythical boundaries of PRTs, enjoy perusing the history of the technological requirements of coupling. But since his informants do not question the power of these things, in his own analysis PRTs and couplings will play the role of what is "in the air." Everybody breathes it in equal proportions. It creates no distinctions. None of the small bifurcations that will turn out to explain the project can be dependent on that vast background common to all projects. The infrastructure, even in the final instance, does not explain the fragile superstructure of the Aramis vehicles. If that indifference to the general "framework" is shocking, let's say that our sociology prefers a local history whose framework is defined by the actors and not by the investigator. Our local history will talk about Aramis, not about guided transportation, mechanical couplings, or monopolistic state capitalism. On the other hand, it will let the actors add whatever they choose to the framework; it will let them take it as far as they care to go.

[INTERVIEW EXCERPTS]

Still speaking with M. Parlat, on the boulevard Victor:

"I still don't understand very well. Why not make cars that stay far enough apart? There'd be no need for nonmaterial couplings."

"Because then you wouldn't be able to handle enough passengers. Each vehicle is small; all passengers are seated. If you wait between cars, it's all over—you'll be processing just a few passengers per hour. You need trains. That's the constraint you start with, from the beginning of PRTs." [no. 3]

Senate hearings, Washington, D.C., around 1965:

Senator Don MacKenzie: *"But Professor, before you do away with private cars with a stroke of your pen, can you show us how you expect*

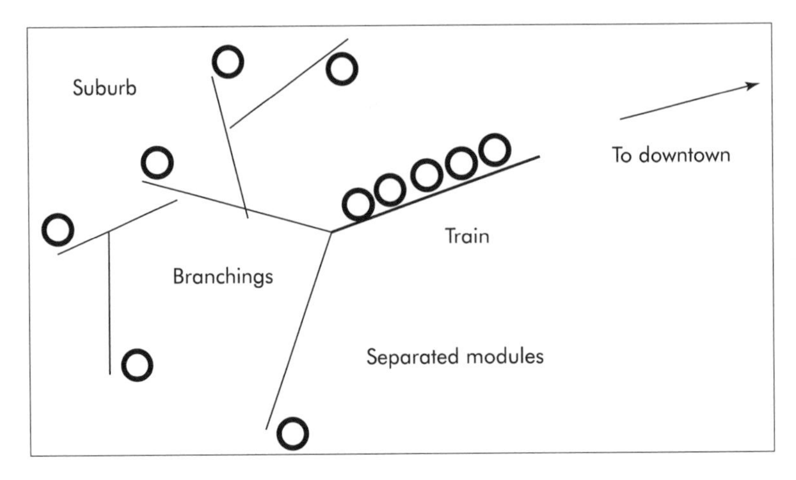

Figure 2. Principle of PRT systems.

to service the suburbs with your PRTs? In the inner city, okay. But just think: What about Los Angeles?"

Jim Johnson, engineer: "*On the contrary, Sir, it's the ideal system for serving large, thinly populated suburbs. What's most expensive in guided transportation, whether you're talking about tramways, subways, or something else? The infrastructure, of course, first and foremost. But then what? The trains, the empty trains that never seem to get calibrated. If you introduce a branch line, either you double the number of trains so as to maintain a constant frequency—and that's expensive—or else you cut the frequency in half. If there's just one branch line, you can do it. But what if there are four, or eight, or sixteen? At the outer edges of the network, there'll be just one train a day—it'll be like the Great Plains in the nineteenth century! And suburbanites will buy a second car. It's inevitable. What you have to do is cut the branching trains into the smallest possible units. Just look at the diagram [Figure 2]:*

When some old lady—a housewife, let's say—wants to go downtown, she fiddles with her keyboard. The computer calculates the best route. It says, 'I'll be there in two minutes'; it's like a taxi. But it's a collective taxi, with no driver, and it's guided by computer. When it arrives, the old lady finds it's carrying a few of her cronies whom the computer has decided to put in the same cab. There's no need for a second car. There's less pollution. And we're still talking about the suburbs, without a heavy infrastructure. It's just like a car."

Senator Tom "Network" Hughes: "*But what do you do about the load factor, Professor? Your isolated train cars, operating far apart, are fine at the end of the line where few people live, but when you get closer to downtown they're going to clog up. It'll be much too slow."*

Jim Johnson: *"Right, right, that's exactly the idea of the train. We put cars together, like a real train, with independent cars, and that way we take care of the passenger load."*

Senator Howarth: *"You're hardly going to couple and uncouple them by hand a hundred times a day!"*

Jim Johnson: *"No, Sir, no, that would be too slow; we're looking for a practical way of coupling. By computer. But we haven't quite perfected it yet, I have to admit."*

Senator Wallace: *"If I may say so, there's something else that hasn't been perfected in this business. What if instead of finding her 'cronies,' as you put it, in this closed car with no driver, your housewife runs into a couple of thugs? (I didn't say 'blacks'—be sure to get that straight.) Then what does she do? What happens to her then?"*

Jim Johnson (at a loss for words): *"Uh . . ."*

Senator Wallace: *"Well, I'll tell you what happens, she gets raped! And the rapist has all the time in the world, in this automated shell of yours with no doors and no windows. You know what you've invented? You've invented the rape wagon!"*

[Shouting, commotion]

[INTERVIEW EXCERPTS]

Liévin, at INRETS: "Aramis? It was the World's Fair. Without that, you can't understand a thing about the project. It certainly wouldn't have started up again in 1981, 1982." [no. 15]

Etienne, at Matra: "And then there was the World's Fair project. That's what got it going again." [no. 21]

Girard: "What explains my 'conversion,' if you like, was the project for the 1989 World's Fair. Every World's Fair presupposes a new form of transportation. Within the range of projects presented, Aramis was truly innovative: France was really going to be able to present something that symbolized French technology at the end of the 1980s. That's what made me change my mind." [no. 18, p. 6]

"If Aramis had been ready in time for the World's Fair, would it have gotten everyone's attention?" I asked.

"Yes, everyone's—it was really a compelling idea."

"But there was no World's Fair, as it turned out."

"Well, no, Chirac didn't want one; he didn't want to upset Parisians with reminders of the Revolution."*

Reuters, September 10, 1989, from our special correspondent Bernard Joerges. *Every World's Fair refurbishes the image of public transportation to some extent. The one that marked the bicentennial of the French Revolution in banner-bedecked Paris, the one that has just closed its doors after a grand ceremony on the Champ-de-Mars, was no exception to the rule. From this standpoint, one of the key features of the fair was unquestionably the completely automated and completely modular transportation system called Aramis. More than motorboats on the Seine, more than moving sidewalks, Aramis is a revolutionary transportation system conceived and constructed by the Matra Transport company, which has demonstrated its technological superiority once again. Specialists in space technology and sophisticated weaponry, the Matra people are shaking up the field of urban transportation, which has been mired in tradition for so long. The Régie Autonome des Transports Parisiens (RATP), responsible for implementing Aramis, has leapfrogged into the twenty-first century thanks to this astonishing display. A train arrives at the station. Of course, as in the VAL system that operates in Lille, or Morgantown's small system in the United States, or Atlanta's, there is no conductor. Elegant little cabs, as cozy and comfortable as a Renault Espace, hold up to twenty visitors each. But here is the surprise: each car is separate. Nothing visible links it to the ones behind: no coupling, no cable, no wire, no linkage of any sort. And yet the cars form a train; they approach one another and merge ever so gently. They stay together as*

*Jacques Chirac, in addition to serving as prime minister from 1986 to 1988, was the mayor of Paris from 1977 to 1995. World's Fairs have been held many times in Paris from 1889 to 1937 and have shaped a number of its landmarks, including the Eiffel Tower. For the new socialist government, hosting a World's Fair in Paris in 1989 seemed a natural way to celebrate the bicentennial of the French Revolution. Obviously, however, the mayor of Paris—who was also the leader of the opposition—would have had to agree.

if by magic. An electronic calculation attaches them together more solidly than any cable. This is what the project engineers call "nonmaterial coupling." Sometimes a slight jolt, a little bump, is felt when two cars come into contact. The most violent shock is psychological—the one that awaits visitors at branching points. Their car pulls away from the train! While one group of riders is transported to one part of the fair, the rest of the train reconstitutes itself and goes on toward another area. No one needs to change trains! No more transfers between lines! Matra and the RATP have invented the transportation system of the twenty-first century, as intimate and personalized as a taxi, as secure and inexpensive as collective transportation. The automobile becomes communal property. Several years before the Japanese and the Americans, while we are still at the stage of trying to make our own Cabinentaxis work, France has been able to get a toehold in a promising market thanks to the World's Fair. People in the transportation industry are simply wondering how much this little marvel must have cost. There is talk of two billion francs! After the Concorde, La Villette, the Rafale, and the nuclear power program, we are well aware that French engineers do not worry about the price tag. True, the fair makes it possibleto justify any extravagance. The revolution (of public transportation) within the Revolution (the French one) is beyond price . . .

By definition, a technological project is a fiction, since at the outset it does not exist, and there is no way it can exist yet because it is in the project phase.

This tautology frees the analysis of technologies from the burden that weighs on analysis of the sciences. As accustomed as we have become to the idea of a science that "constructs," "fashions," or "produces" its objects, the fact still remains that, after all the controversies, the sciences seem to have discovered a world that came into being without men and without sciences. Galileo may have constructed the phases of Venus, but once that construction was complete her phases appeared to have been "always already present." The fabricated fact has become the accomplished fact, the *fait accompli.* Diesel did not construct his engine any more than Galileo built his planet. Some will contend that the engine is out of Diesel's control as much as Venus was out of Galileo's; even so, no one would dare assert that the Diesel engine "was always already there, even before it was discovered." No one is a Platonist where technology is concerned—except for very primitive, basic gestures like the ones Leroi-Gourhan calls "technological trends."

This rejection of Platonism gives greater freedom to the observer of machines than to the observer of facts. The big problems of realism and relativism do not bother him. He is free to study engineers who are creating fictions, since fiction, the projection of a state of technology from five or fifty years in the future to a time *T*, is part of their job. They invent a means of transportation that does not exist, paper passengers, opportunities that have to be created, places to be designed (often from scratch), component industries, technological revolutions. They're novelists. With just one difference: their project—which is at first indistinguishable from a novel—will gradually veer in one direction or another. Either it will remain a project in the file drawers (and its text is often less amusing to read than that of a novel) or else it will be transformed into an object.

In the beginning, there is no distinction between projects and objects. The two circulate from office to office in the form of paper, plans, departmental memos, speeches, scale models, and occasional synopses. Here we're in the realm of signs, language, texts. In the end, people, after they leave their offices, are the ones who circulate inside the object. A Copernican revolution. A gulf opens up between the world of signs and the world of things. The R-312 is no longer a novel that carries me away in transports of delight; it's a bus that transports me away from the boulevard Saint-Michel. The observer of technologies has to be very careful not to differentiate too hastily between signs and things, between projects and objects, between fiction and reality, between a novel about feelings and what is inscribed in the nature of things. In fact, the engineers the observer is studying pass *progressively* from one of these sets to another. The R-312 was a text; now it's a thing. Once a carcass, it will eventually revert to the carcass state. Aramis was a text; it came close to becoming, it nearly became, it might have become, an object, an institution, a means of transportation in Paris. In the archives, it turns back into a text, a technological fiction. The capacity of a text to weigh itself down with reality, or, on the contrary, to lighten its load of reality, is what endows fictional technologies with a beauty that the novel we've inherited from the nineteenth century has difficulty manifesting nowadays. Only a fiction that *gains or loses reality* can do justice to the engineers, those great despised figures of culture and history. A fiction with "variable geometry": this is what needs to be invented, if we are to track the variations of a technological project that has the potential to become an object.

"Personal Rapid Transit systems, nonmaterial couplings, the composition of trains—all this is beginning to take shape," Norbert told me. "Now let's try to see whether we can pin down the archaeology of the project, the earliest ideas, the creative spark. Often the initial idea doesn't count for much in a project, but my hunch is that this time it must have played a role."

[INTERVIEW EXCERPTS]

At the Conseil Général des Ponts et Chaussées, M. Petit sits in a large office. He is speaking very rapidly, obliged by our questions to return to a past he finds very remote.

"Ah, Aramis. In the beginning it wasn't Aramis, we didn't even have cars, we had *programmed seats*. Yes, that's how we started. I was with DATAR at the time. DATAR was a powerhouse then; they had a lot of money, and all the ministries had to pay attention to them. DATAR, you know, was Olivier Guichard, under de Gaulle; it was 'joint development of the Sahara regions.' Our bad luck: we lost the Sahara when we pulled out of Algeria. We got medals embossed with camels and palm trees. Guichard didn't give up. He created DATAR; it was his idea. Directly attached to the prime minister. Development in the Sahara, in France—it's pretty much the same thing.

"Well, the highway system was a mess, split up among several ministries. So we produced an overall highway management plan. Okay, we said, no point in France should be more than two hours away from any other point. Whatever means of transportation is used. Railways, iron on iron, you know, it's not that great. As soon as you go fast, you lose your contact. In fifty years there won't be any more trains. We needed something in the range of 300, 400 kilometers an hour.

"Bertin came to see us. 'The future is in the air cushion.' Yes, the aerotrain—that was us. We built a line in Orléans. You can still see it. Well, it didn't catch on. We gave it to the SNCF, which shut it down in a hurry. But for what amounted then to 50 million francs, I shook up the SNCF. It was a gift. The high-speed train (TGV) is the bastard child of the aerotrain. They turned their 12,000 engineers loose so the aerotrain would never happen, and they came up with the TGV!

"Okay, but there was another hole in public transportation—the metro. There's nothing you can do; you can't go beyond 18 kilometers an hour, assuming you have stations every 400 meters and a maximum acceleration of 1.2 m/sec^2. Beyond that, it musses your hairdo; it shakes people up. But wait, we said, maybe there's something better than the metro. With moving sidewalks, you can't accelerate faster than 3 m/sec^2. If folks are walking on the sidewalk at 6 kilometers per hour, when they come to the end they're catapulted. That doesn't work.

"You know, when you invent an urban transportation system, you always get into trouble with the little old blind lady with a heart condition who gets her umbrella stuck. You always have to take her into account.

"Then I had kind of a crazy idea. I said to myself that there were people in factories who made transfer machines. You know, machines that take anything—say, bottles—and zap, give them infinite acceleration, from 0 to 20 kilometers an hour, instantly. Whether you're talking about fragile bottles or little old ladies, it's the same sort of problem. I thought about munitions factories. You can't let the cartridges explode, yet people have to be able to pick them up and put them down.

"I asked the army. They said: 'That's kinematics, and kinematics is Bardet.' Gérard Bardet was synonymous with the company he had founded, Automatisme et Technique. The only one in France, the only one in Europe I think. He had just won a competition for Winchester cartridges. He was filling them with powder, with 600 leads and all that, at 25 cartridges a second. I called him up.

"A very appealing guy. He'd had a hard life, lots of upheavals. He set up his society as a cooperative so as to give it to his employees. You don't see that very often.

"Okay, so I asked him the question. How do you transport big loads, around 100,000 passengers an hour? He said: 'Let's go see what our mad inventors have in their back yards.'

"You've no idea! The word got around: 'If DATAR is helping scientists, that's great!' I had all sorts of mad inventors trooping through my office. One of my buddies from the Ecole Polytechnique even dragged me to his house. There wasn't a stick of furniture left. On the ceiling, there was a vacuum cleaner on rails. 'Bertin gets down on the floor and blows,' he explained. 'I get on the

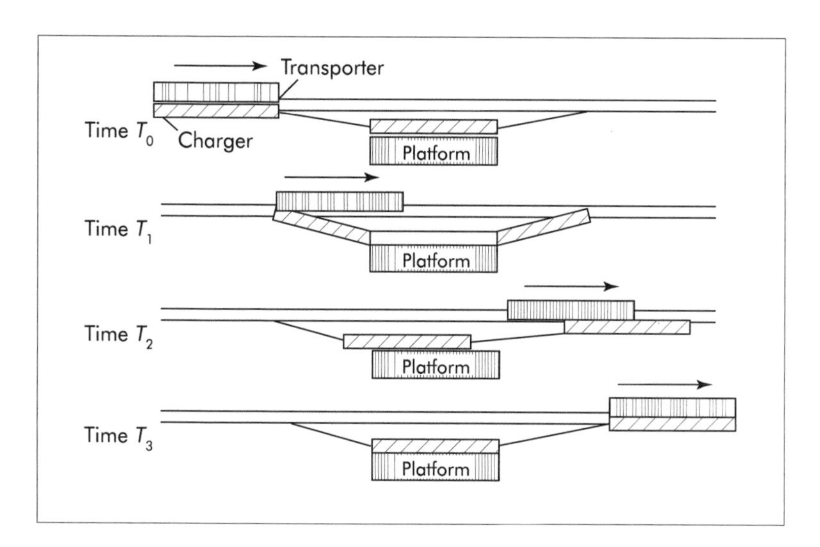

Figure 3. The AT-2000 train.

ceiling and breathe in. The inverse of an invention is still an invention. Does that interest you?'

"Bardet said, 'This isn't getting us anywhere. Let's make an invention matrix.' Nine boxes by nine. He put in every form of transportation you can think of. A chart worthy of Mendeleev. We invented terrific subway systems. In one box of the matrix we noticed, for example, that what's dumb about subway trains is that they stop at stations. On the other hand, a subway that doesn't stop . . .

"So what can we do? Well, a transfer machine. You cut the subway train in half, lengthwise. You always have one branch of it at the station. Another branch charges into the tunnel without stopping. Near a station, those who want to get off move into the corridor. The doors close, it's uncoupled, it slows down. Meanwhile, the people who want to get on have gotten into the corridor-branch that was in the station. They speed up and rejoin the branch that didn't stop. *[He draws a hasty sketch on a notepad—see Figure 3.]*

"And we went on like that. We got up to incredible volumes, 100,000 to 200,000 passengers an hour. The mockup we did cost DATAR 30 million francs.

"The computer was full-scale. The mockup was in all the fairs. It was called the AT-2000. I was even on television with Alexandre Tarta; the tape must still be around somewhere."

"And what about Aramis, M. Petit?"

"Well, Aramis was the eighty-first box of the invention matrix, the niftiest of all, it was the programmed metro seat. The traveler merely goes to the station. He sits down, punches in the program, and opens up his newspaper. When the thing stops, he looks up, puts away his paper, and there he is, where he wanted to go. It's point-to-point, with no connections, no stops at intermediate stations. The eighty-first box was the most seductive of all for a with-it technocrat eager to impress a client.

"Meanwhile, Matra had a whole lot of ideas. They wanted to diversify. They were involved in military business, which let Lagardère make a good show. They hit it off with Bardet. [no. 40]

The difference between dreams and reality is variable.

The guy who spray-paints his innermost feelings on the white walls of the Pigalle metro station may be rebelling against the drab reality of the stations, the cars, the tracks, and the surveillance cameras. His dreams seem to him to be infinitely remote from the harsh truth of the stations, and that's why he signs his name in rage on the white ceramic tiles. The chief engineer who dreams of a speedier metro likewise crosses out plans according to his moods. But if the AT-2000 had been developed, his dream would have become the other's world. The spray-painting hoodlum would then be living partly within the other's dream brought to life, just as he is living in the waking dream of Fulgence Bienvenüe. In Paris, a war of the worlds is raging, a war of dreams, a war quite different from the opposition between states of feeling and states of affairs, between soft subjects and hard technologies. Dreams seeking to be realized are shaping Paris, working through its subterranean spaces and stations. They touch and try one another. The subway is too slow; it can be redone. The engineer Bardet is no less impatient than the hoodlum. He, too, wants to change the metro, to change life. Let's be careful not to oppose cold calculators to hot agitators. Neither is more spontaneous than the other. Petit is influenced by the Americans' PRT, which is hardly surprising; then again—and this is much harder to believe—the illiterate hoodlum does his tagging spontaneously in English and in the graphic style of the New York City gangs!

As for Bardet, he's dreaming too. For where, if not in a dream, could one compare a 130-pound grandmother headed for the Sacré-Coeur station with a 100-gram cartridge that a transfer machine picks up on an assembly line? You said "transfer"? Well, well! Could the unconscious be full of machines as well as affects? The entire Paris metro system—in fact, all the transportation systems of the world—find themselves brought together in an eighty-one-box chart on DATAR's table. Dreams change the scale of phenomena, as we know: they allow new combinations and they mix up properties. So: an engineer's dream?

"Well, my dear Watson, what do you think? It is all perfectly clear?"

"Certainly," I said, a bit uneasy to be feeling so sure of myself. "Of course, I don't know much about it, but Aramis is an engineer's toy, one of those far-fetched ideas that didn't grow out of a needs analysis. That much is obvious right away."

"Wrong, as usual," Norbert replied amiably. "On the contrary, it was to avoid far-fetched inventions like those the Lépine competition produces that Petit and Bardet drew up their matrix. I questioned the director of SOFRETU, and he confirmed Petit's account point for point."

From a grimy little notebook, Norbert extracted his interview notes.

[INTERVIEW EXCERPTS]

The director of SOFRETU:

"I'm not starting with inventions or components; no more brilliant and unworkable ideas. I'm starting with passengers, with their real needs, with uses. The ideal, for passengers, is what? It's not to think, not to slow down, not to stop, not to transfer, and to arrive at their destination nevertheless. That's point-to-point transportation. That's Aramis." [no. 18]

"I even came across a document by Bardet from 1969 or 1970. You know what he says?"

[DOCUMENT: FROM AUTOMATISME ET TECHNIQUE, REPORT ENTITLED "LES TRANSPORTS URBAINS," TECHNICAL NOTE; EMPHASIS ADDED]

Continuous transportation is the possibility of adapting later *to any evolution, no matter how unpredictable, of technology* or of urban planning; in other words, it amounts to *respect for the indeterminacy of the future.* This safeguarding of the future has to be envisaged both on the technological level and from the standpoint of serving constantly evolving metropolitan populations. On the technological level, it is important to stress that *no hypothesis was made at the outset* regarding the technologies to be used . . . Systems of continuous transportation are essentially based on a kinematic principle and will always be able to incorporate future technological projects.

"You see," my professor continued, "it's more complicated than you think . . . It's the opposite of an engineer's idea: it's a system-idea, open to the unpredictable. No, no, we're off to a good start. A textbook case, my friend, a real textbook case. 'Respect for indeterminacy,' that's what we teach our students; to start from principles, needs, systems and not from technology. It's really rotten luck." Then, looking at his watch, he added: "We're going to be late for our meeting on the quai des Grands-Augustins."

We got off the bus, which was stuck in a traffic jam, and we studied the subway map, trying to calculate which route would involve the fewest transfers.

"Aramis ought to be making this calculation," I said.

"Exactly, and we wouldn't have to transfer at Châtelet. You see, my friend, Aramis really is an idea for consumers, not an engineer's idea. It's the one time they were actually thinking about us—and it didn't work."

"You mean, 'The one time they were thinking about our not having to think about anything,'" was my clever riposte.

"Engineers dream, but they're not crazy," my mentor replied

primly, without acknowledging my cleverness. “What does Bardet produce, Mister Oh-So-Reasonable Young Engineer? A critique of the urban society of his time. Well? Does it surprise you that a kinematician should get involved in making a whole movie script out of cars, happiness, and the future of civilization?”

[DOCUMENT: REPORT BY AUTOMATISME ET TECHNIQUE, 1969 OR 1970]

To summarize, without getting bogged down in a purely sterile critique, let us note [in 1970] that the situation is triply paradoxical:

—While the automobile still seems to be the fastest (though costliest) solution for urban transportation in the short run, its very proliferation will increasingly cut down on its speed, which will soon become unacceptably slow; at the same time, automobiles will increase to dangerous levels the atmospheric pollution that they inevitably produce. This is the paradox of the scientific organization of total asphyxia—in the broadest sense of the term.

—At a time when efficiency has the status of dogma, we are all subject to its discipline, and in our stressed-out state, before and after work, we all have to put up with physically exhausting compressions in uncomfortable spaces and annoying waiting periods owing to breakdowns in the traffic flow. This is the paradox of antisocial behavior in a society that would like to see itself as social.

—Finally and in more general terms, isn't it unreasonable that in this speeded-up century the time it takes us to cover the distance between home and the airport hasn't changed, coming or going? This is the paradox of "constant time," whatever the distance covered.

In the face of these observations, which are not just ploys in some amusing mental game but have social repercussions whose economic consequences weigh heavily on us, *is technology powerless* [p. 7]?

Automatisme et Technique doesn't think so. For the past three years, with the cooperation of public agencies that are especially concerned—the RATP on the one hand, DATAR on the other—we have been working on theoretical research projects and technological developments leading to new solutions characterized as much by their performance as by their variety and adaptive flexibility. [p. 8]

The key to this innovation is a kinematic principle. Public transportation has to be considered a particular case of *continuous transportation.*

The application of "continuous kinematics" to transportation problems makes it possible, above and beyond the possibilities of classic transportation systems, to *reconcile* research aimed at greatly increased speeds and heightened comfort with concern for maximizing fine-tuned service.

In order to *translate* this objective, Automatisme et Technique has spelled out two rules that apply to the traffic flow:

—Passengers must be able to pass through the intermediate stations on their itinerary without stopping.

—An increase in the number of stations along a connecting line must not affect either the speed or the volume of service on the line.

The first consequence of these rules is that they lead to *dissociation* of the "transportation" function as such from the function of "access" to the transport mechanism, whereas in classic transportation systems these functions are taken care of by a single mechanism. [p. 13].

No technological project is technological first and foremost.

"What's that engineer poking his nose into?" you may well ask. "Why is he criticizing society, pursuing his own politics, his own urban planning? An engineer answers questions, he doesn't ask them." This is the image of engineers held by people who think technology is neutral, or (it comes down to the same thing) that technology is purely a means to an end, or (and this still amounts to the same thing) that the only goal of technology is technology itself and its own further development. Bardet, as we have seen, defines his goals and questions for himself, even if he is defensive about playing "amusing mental games" or making "sterile critiques." He's a sociologist as well as a technician. Let's say that he's a sociotechnician, and that he relies on a particular form of ingenuity, *heterogeneous engineering,* which leads him to blend together major social questions concerning the spirit of the age or the century and "properly" technological questions in a single discourse.

How does this blend come about? Not by chance, but by a precise operation of *translation.* Urban transportation systems are being asphyxiated, Bardet says; this asphyxiation, as he sees it, is contrary to the spirit of the age. This intolerable situation has to end. How can we put a stop to it? Kinematics deals with continuous transportation of bottles, cartridges, or jam jars. And who controls kinematics? Bardet and his company. Between the asphyxiated society of automobiles and transfer machines in factories, there is no connection whatever. Bardet, approached by Petit, is going to make this connection. The price to pay is an innovation: the discontinuous transportation of people, which no one knows how to improve, has to be viewed as a particular case of the continuous transportation of things, which Bardet knows how to improve. The result? A chain of translation: there is no solution to the problems of the city without innovations in transportation, no innovation in transportation without kinematics, no kinematics without Automatisme et Technique; and, of course, no Automatisme et Technique without Bardet.

People always wonder how a laboratory, or a science, can have any impact at all on society, or how an innovation arises in the mind of its inventors. The answer is always to be found in the chains of translation that transform a global problem (the city, the century) into a local problem (kinematics, continuous transportation) through a series of intermediaries that are not "logical" in the formal sense of the term, but that oblige those, like DATAR, who are interested in the global problem to become interested, through almost imperceptible shifts, in the local solution. The innovation, as Bardet says, will make it possible to "translate" and to "reconcile" contraries in order to establish chains of translation and to situate Bardet's expertise as the *obligatory passage point* that will resolve the great problems of the age. The work of *generating interest* consists in constructing these long chains of reasons that are irresistible, even though their logical form may be debatable. If you want to save the city, save Bardet. This implication is not logically correct, but it is socio-logically accurate.

I was outraged by what my professor, with a certain satisfaction, was calling "chains of translation." He seemed to take great pleasure in seeing huge interests drift off toward little laboratories. For my part, I was deeply shocked by that sordid, self-interested vision of the engineer's work.

"It's just a way of talking about priming the pump, if I understand correctly," I said with more feeling than my professor usually permitted (it was never a good idea to let him think one was naïve). "Bardet is making money by making silk purses out of sows' ears. He's a cynic. But what would a real engineer have done in his place?"

"My dear young friend, I forbid you to speak, or even to think, ill of Bardet. He's a great engineer, a real one. You're always jumping from one extreme to another. You show up here convinced that technology is neutral and beyond question. You get your nose rubbed in a project—for your own good—and you conclude that it's all a matter of pork barrels and white elephants. You move too fast. You really do have a lot to learn. An engineer has to stimulate *interest:* that's the long and the short of it. And he also has to convince; that's the Law and the Gospel. You can't put any real engineer 'in Bardet's place' (as you term it) except a bad one, some imbecile who doesn't interest and doesn't convince and whose kinematics has never gotten beyond the end of its transfer function.

"In contrast, look at the beauty of Aramis and PRTs. It's a fantastic invention. To discourage residents from taking their cars, you merge cars with public transportation. There's only one way you can do that: you have to get people to see public transportation the way they do automobiles, so they'll take public transportation instead of their own cars. It's a matter of mimicry, just like in the jungle. The worn-down citydweller stops distinguishing between his private car and his Aramis car. He literally takes one for the other! Let's give collective transportation some of the automobile's most interesting features—point-to-point service, no transfers, comfort, intimacy—plus all the advantages of public transportation: speed, train service that copes with the traffic flow, low cost (to the user), lack of responsibility (again, for the user). No one will want to do without it. Just look at these great interest curves: DATAR, the RATP, Paris, before you know it the whole world. Yet it's still contemporary twenty years later! The diagnostic hasn't budged: everything has only gotten worse, in cities. And in the center, in what has become the center, resting at the heart of the mechanism: kinematics, continuous transportation. A peaceful revolution desired by all. And you'll never again have to change trains or wash your car. Cars for everybody. No, the importance of Bardet's innovations can't be overestimated. I don't like the word, but Bardet is a genius. Unfortunately, we can't interview him. I've talked to his wife: he's very old, and too ill to answer any questions."

"Still, he blew it," I thought. But I kept my opinion to myself.

Justice and young engineers with no memory are hard on projects that fail.

The ultimate defect of projects—they die—takes us back to their beginnings: they were condemned from the start because some crazy engineers had mistaken dreams for reality. The verdict is clear: Personal Rapid Transport systems died because they were not viable. But biological metaphors are as dangerous for technological organizations as they are for living organisms. You can't say that PRTs died because they weren't viable, any more than you can say that dinosaurs, after surviving for millions of years, died out because they were doomed or ill-conceived. Aramis died—in 1987—and its accusers claim that it was nonviable from the beginning, from 1970. No: Aramis was terminated in 1970, and the explanation makers were kicking a dead horse when they claimed that it hadn't been feasible from the start and that they themselves had been saying so all along. Blessed are the lesson givers, for they will always be right—afterward . . . Don't ask them for immediate opinions on the Concorde, or the future of computers, or aerotrains, or superconductivity, or telephones. You'll get the answers only ten or twenty years later, and they'll say they knew all along that the project was not viable. No, Aramis is feasible, at least as feasible as dinosaurs, for life is a state of uncertainty and risk, of fragile adaptation to a past and present environment that the future cannot judge.

The innovations produced by people like Bardet, Petit, Boeing, Otis, and Daimler-Benz are real, important, and exciting. What is at stake, owing to the fusion between the worlds of continous kinematics and public transportation, is a *compromise.* Innovation always comes from a blending or redistribution of properties that previously had been dispersed. Prior to the fusion of kinematics and public transportation, no one had noticed that the *transport* function could be separated from the *access* function. This distinction is what allows the technological compromise to emerge: let's invent a system that never slows down and that nevertheless allows for personalized access. Aramis is a textbook case. No one in his or her right mind can be opposed to a PRT that marries, fuses, blends the private car with public transportation, a project that saves us from asphyxiation. No one can criticize the management of a project that leaves the future open, that does not make premature judgments about the technological components. Neither wicked capitalists nor purveyors of useless gadgets are the driving force behind this effort. No, it is a matter of real inventions designed to meet real needs proposed by real public servants and supported by real scientists. A dream, yes, a dream. In any case, it is paved with good intentions.

"Always assume that people are right, even if you have to stretch the point a bit. A simple rule, my dear pupil, when you're studying a project. You put yourself at the peak of enthusiasm, at the apex, the point when the thing is irresistible, when what you really want, yourself, is to take out your checkbook so you can, I don't know . . ."

"Buy a share in the Chunnel?"

"That's it, or even shares in the Concorde."

"Even in La Villette?"

"Which one, the first scandal or the second?"

"The second."

"Oh, the La Villette museum. I don't know; it's a disaster, but after all, why not, it had to be tried. Never say it's stupid. Say: If I were in their shoes, I'd have done the same thing."

"Even in that business of the sniffer planes?"

"Of course, silly boy, you would have bought into it, and not because you're naïve; on the contrary, precisely because you're a clever fellow. It's like the Galileo affair. You have to get inside it until you're sure: that one is guilty; he should be exiled, and even, yes, even fried a little, the tips of his toes at least. Otherwise, if you think differently, you're a little snot. You play the sly one at the expense of history. You play the wise old owl."

"The one that always arrives at nightfall, like the cavalry?"

"Ah, I see they do teach you something, after all, in Telecommunications. Yes, you have to reread your Hegel because, you know, technological reality isn't rational, and it's no good rationalizing it after the fact."

In the list of books to read, Hegel came after *F,* for Favret-Saada, and *G,* for Garfunkel or Garfinkel, like the singer but not so easy to set to music.

[INTERVIEW EXCERPTS]

At RATP headquarters, on the second floor where the directors have their plush lairs, M. Maire is sitting in his office.

"Bardet was a really nice guy. I remember in 1968 or '69 he invited us for lunch in a little bistrot—Girard, Antoine, and me. He'd invented the AT-2000. He told us he was worried, he didn't understand why people were skeptical about the AT. 'Why aren't you supporting the AT-2000?' We told him we didn't think it was very reliable. I don't remember whose idea it was, during lunch, to try slicing the trains in the other direction, crosswise. That's it right there—that's where Aramis came from. He applied for the patent on it a few days later.

"He'd also invented the modular train. *[Sketching on a scrap of paper—see Figure 4.]* That wasn't stupid, because it did away with the need for side stations. Part of the train *[modules c and d]* didn't stop, and hooked up with the front section *[modules x and y],* which had moved out of the station. Before reaching the station, that section shed its rear compartment *[modules a and b].* So there were always cars at the center of the train that didn't stop. The problem was that passengers had to move to the rear cars, which were the only ones that stopped at stations. If there were a lot of stations, there would have been quite a lot of movement.

"So Bardet applied for a patent on trains consisting of small programmed vehicles. 'Small vehicles,' since they were to be comfortable and intimate, all

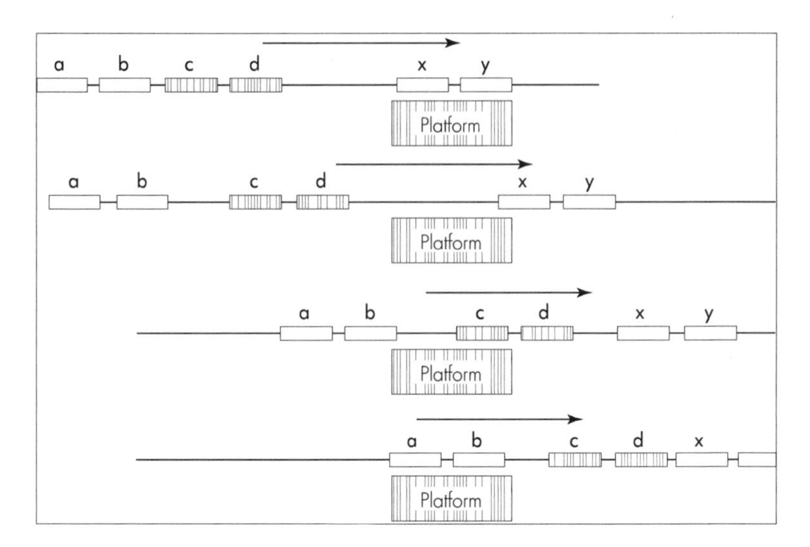

Figure 4.

going in the same direction and easy to insert in cities without heavy infrastructures. 'Trains,' because moving together train-fashion is the only way to ensure adequate flow. 'Programmed,' so passengers would only have to punch in their destination on the dashboard and the vehicle would head straight to the desired destination. Then Matra bought Bardet's patents *[he sketches Aramis—see Figure 5]*. [no. 22]

"Since the witness has moved from DATAR to Matra, we have to move from the public to the private sphere as well, to gather our testimony. See, it's not that hard. As soon as somebody's name is mentioned, you call him up, you make an appointment, and you go see him."

"Do you always get a good reception?"

"Always, and the more important the people are, the less they keep you waiting."

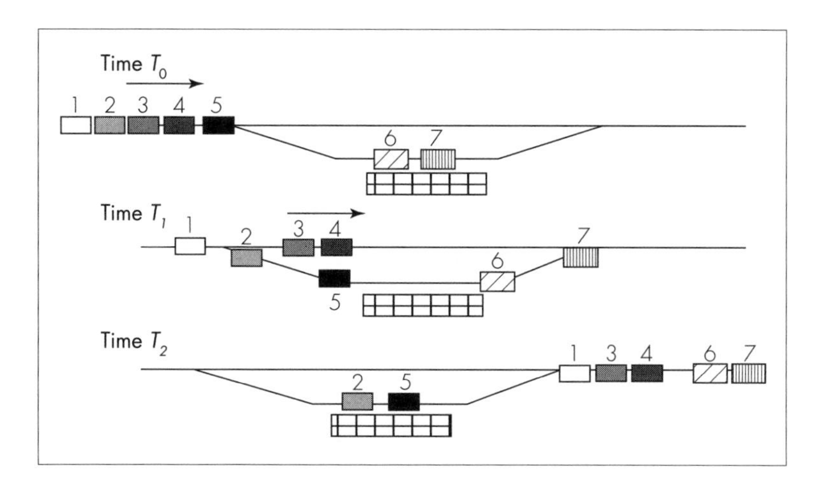

Figure 5. Train consisting of small programmed vehicles.

[INTERVIEW EXCERPTS]

The scene is Matra Transport's suburban headquarters, in a building decorated in the inevitable postmodern white tiles. At the end of the slick, white main hall, you can see the splendid white casing of Aramis. The director, M. Etienne, is speaking:

"Lagardère had put Pierre Quétard in charge of diversifying Matra in the civilian sector. He took a good look all around. There were some pluses, and some mistakes. That was normal.

"Anyway, Quétard was on the lookout. He had been to see more or less everybody, offering Matra's services, insisting on their advanced technological competence, and also on the logic of the complex systems that were among their specialities . . . Petit sought out Matra—or Quétard sought out Petit, I don't remember now. Anyway, Petit said: 'We're onto something terrific here; I'm ready to put money into it. There's this little company that's in over its head. You've got to work with them.'

"He'd even come to see me at the DTT (the bureau of ground transportation), but I don't think I gave him any money at that point. You have to remember that DATAR in those days wasn't what it is now. It was a powerhouse. It had been set up so it could really do something. For the ministries, a *nyet* from DATAR was a real catastrophe, at the time.

"Well, DATAR was obsessed with the growth of the Paris region, and it was trying to support public transportation. It was interested in creating a new intermediate urban network. So new forms of public transportation were enormously intriguing. In the long run, their proactive approach accomplished some things, too; that was what was behind the new cities. They're still here. That was also the force that called a halt to the growth of Paris . . .

"In any event, Matra came to an understanding with Bardet. Matra didn't really 'purchase' Bardet's patents. Or at least it was more complicated than a purchase. The contract is hard to analyze. In fact, except for a core of essential ideas that stayed pretty much the same, all the rest was Matra's doing, rather than Automatisme et Technique's. Even the name is ours." [no. 21]

"So we finally know why it's called Aramis?"

"Yes, Matra gave Bardet's little programmed vehicles the bizarre name Agencement en Rames Automatisées de Modules Indépendants dans les Stations, meaning 'arrangement in automated trains of independent modules in stations.' Aramis, for short. It has a nice ring to it, 'Aramis.'"

"The name is different, but when you get right down to it, if you read the documents from that period, Matra is making the same arguments as Petit and Bardet."

[DOCUMENT: FROM "ENGINS MATRA," A REPORT ON ARAMIS, UNDATED BUT PROBABLY FROM 1971]

The automobile marks our generation. Weekend gridlock and urban pollution are upsetting, but they don't stop its development. The quality of service it offers—speed, availability, suitability for door-to-door transportation—is incomparable, and accounts for its appeal. Aramis, a system of urban and suburban on-site transportation, offers an alternative to the automobile, whose very proliferation cuts down significantly on its performance [p. 1] . . . Aramis does not stand in competition with the automobile, but as a *complement* to it. By offering users a free choice between two equally attractive methods, *it gives the automobile's "prisoners" their freedom back*. By pulling part of the traffic off the roads, it improves traffic conditions . . . Aramis' users constitute a clientele that appreciates the advantages of the automobile, while rejecting its disadvantages. For Aramis is like the automobile: it offers comfort, availability,* the absence of interruptions. But *in addition,* it offers speed (50 km/h), safety, punctuality. An electric-powered system on pneumatic wheels, Aramis also protects the environment (no atmospheric pollution, no noise) [p. 13] . . . To choose Aramis today is to win the wager already, the one our children will make tomorrow in order to live in cities that have a human face.

"The style is better than Bardet's."

"Yes, and you'll also notice that they cast a wider net. A lot more people are interested, or might become interested. It's no longer just the State that is presumed to be interested in Aramis for the purpose of improving the infrastructure or serving the public. Now it's drivers themselves, 'prisoners' freed from their chains, who are achieving their goals—by way of Aramis. Not bad! Notice that for the first time the *market* is making its appearance in the form (a somewhat curious form, I admit) of the consumers' desire to buy cars but to use Aramis as well, so as to cut down on automobile traffic. You see, it was a good idea to move from the public to the private sphere. It's always crucial to get hold of the original documents."

The "market forces" of the private sector are actors like the others.

The analysis of technological projects often runs aground because the observers are intimidated by the economic forces that, like the technological determinism we saw earlier, are assumed to go up and down the

*In the industrial world, availability is not a moral virtue comparable to charity; it is a practical virtue which indicates that the machine or the means of transportation in question has not broken down, that it is available. It is usually calculated as a ratio that should be above 0.96, at least for a metro line.

boulevard Saint-Michel with the power of an R-312 bus. Yet consumers are seduced by Aramis just as DATAR is. Consumers, too, are invented, displaced, translated, through fine chains of interest. Bardet and Petit ask DATAR: "You want to save the city? Limit the growth of Paris? Then from now on you have to be interested in kinematics, in transfer machines, and in the AT-2000." "You really want to profit from the advantages of the automobile?" Matra's people ask prospective consumers. "Then you have to climb into a cabin that is almost the same, but guided, called Aramis." In each case you have to make a tiny shift, a nearly imperceptible detour.

Is this process of translation "false," "misleading," "rhetorical," or "illogical"? Does Aramis really meet a need? We don't yet know. It all depends. On what? On what happens next, and on how much you trust the spokespersons of all those needs and interests. Bardet, Petit, and Lagardère are self-designated representatives who speak *in the name of* the city, the future, pollution, and what consumers really want. At this stage there is no difference between Petit, a highly placed government official who speaks in the name of all urban Frenchmen, and the industrialist Lagardère, who speaks in the name of all consumers. Rather than focusing on the artificial difference between State and industry, the public sector and the private sector, let's choose the more refined notion of *spokesperson,* and find out, next, whether the constituent groups turn out to be well represented by those to whom they have given their mandate. The spokespersons assert that automobiles must be supplemented, complemented, by Aramis. They are the ones, too, who claim that all their constituents would say, would think, or would mean the same thing eventually, if only people would go to the trouble of questioning them directly. The representatives surround themselves with unanimity. To hear them, the conclusion seems obvious, irresistible: Aramis has to exist, Aramis can exist, and Matra is the company best positioned to bring it into being. Drivers cannot not want to give up their cars. It isn't a question of bad faith here, or cheating, or engineers getting carried away with themselves. Are they mistaken? We cannot know until they have *explored* the world and *verified* whether the city, cars, the powers that be, pollution, the epoch are following them or not. In preference to cumbersome notions such as market forces or the irresistible thrust of technology, let's choose *assemblies of spokespersons* who bring together, during a single meeting, around a single table, different worlds. The highly placed official speaks in the name of developing the French infrastructure and supports the project of the transportation minister—who speaks in the name of the government, which speaks in the name of the voters. The transportation minister supports Matra's project, and Matra speaks in the name of captive drivers, who support the project of the engineer, who speaks in the name of cutting-edge technology. It is because these people translate all the divergent interests of their constituents, and because they meet together nevertheless, that the Aramis project can gain enough certainty, enough confidence, enough enthusiasm to be transformed from paper to prototype.

[DOCUMENT]

Matra's primary vocation is systems development for military and space applications. Its success stories in cutting-edge industries are well known. In pursuit of its objectives, it benefits from homogeneous multidisciplinary teams like the ones currently applying a tested methodology to new transportation systems. Matra is the very model of an industrial company whose size and dynamic decisionmaking structures are perfectly suited to succeed with a project like the Aramis system. [p. 2]

"And here's one more competence, right in the middle, that serves as an obligatory crossing point," I said with the satisfaction of a pupil who has learned his lesson well.

"Very good; but notice that what's at center stage isn't Bardet's competence any more, it isn't kinematics. It's the high-tech capability of a company that's getting a foothold in public transportation. We're shifting from a specific know-how to a general savoir-faire: system building. Things are beginning to shape up. Two very important new actors are backing Aramis now: a business and a market. No matter that the company is a newcomer and that the consumers in question exist only on paper. Somebody who has the prestige of sophisticated military contracts behind him and who expresses the will of millions of 'captive' car owners can't fail to get everyone's attention, especially the ones who hold the purse strings. Yes, Aramis is too beautiful not to come into being. If it didn't exist, somebody would have to invent it."

"Well, they did invent it. Look at this document."

[DOCUMENT]

On April 13, 1972, Michel Frybourg, director of the Institute for Transportation Research (INRETS), and Jean-Luc Lagardère, president of the Matra Motor Corporation, signed an agreement to construct an Aramis prototype in Orly at a cost of around 5 million [1973] francs.

The actors come in varying sizes; this is the whole problem with innovation.

Before a revolutionary transportation system can be inscribed into the nature of things, the transports of enthusiasm shared by all these revolutionaries, industrialists, scientists, and high officials have to be inscribed on paper. *Verba volent.* The agreement signed on April 13, 1972, by spokespersons for the minister and Matra was intended to establish the financial participation of each party, to define the prototype, to describe the development phases, to specify who would control the results, to decide who would possess Aramis' patents and licenses if it were to come into existence, to agree on how each party would pay its share, and finally, in the case of failure, to determine how each one would bow out with dignity, without trials or litigation. But who are Messieurs Petit and Lagardère? They do not have an essence that has been fixed once and for all. They can speak in everyone's name, or no one's; it all depends. Petit may speak for all French people, or for DATAR, or for one of DATAR's departments, or for a member of one of its departments, or in his own name, either as a transportation specialist or as a private individual. He may speak solely in the name of his own imagination. Someone else, or his own unconscious, may even speak for him. Depending on his relative size, he may capture everyone's attention for ten years, or that of just one person for a mere instant. He may be called Mr. Large or Mr. Small. Here we have an essence so elastic that a single sentence, "Mr. Petit is interested in the project," may be translated into a whole gamut of sentences, from "50 million Frenchmen are solemnly and eternally committed to Aramis," to "His imagination is running away with him, but in a couple of minutes he'll have forgotten the whole thing." Now, this variation in the relative size, in the representativeness of the actors, is not limited to Mr. Petit; it characterizes all members of a technological project. Mr. Lagardère supports the project, to be sure, but who can say whether his stockholders will follow? He, too, varies in relative size. Let him be reduced to minority status by his board, and the enormous actor who had millions of francs to contribute is reduced to the simple opinion of a private person whose interest in Aramis commits only himself and his dog. In a project's history, the suspense derives from the swelling or shrinking of the relative size of the actors.

Although this variability can never be eliminated, its scope can nevertheless be limited. Here is where law comes into its own. No technology without rules, without signatures, without bureaucracies and stamps. Law itself is no different from the world of technologies: it is the set of the modest technologies of writing, registering, verifying, authenticating that makes it possible to line up people and statements. It is a world of flexible technologies coming to the aid of even more flexible technologies of interest in order to allow slightly more solid technologies to harden a bit. A signature on a contract, an endorsement, an agreement stabilizes the relative size of the actors by lending to the provisional definition of alliances the assistance of the law, a law whose weight is enormous because it is entirely formal and because it applies equally to everyone. Mr. Lagardère may vary in size, the ministry will change hands ten times—it would be unwise to count on stability there; but the signatures and stamps remain, offering the alliances a relative durability. *Scripta manent.* That will never be enough, for signed documents can turn back into scraps of paper. Yet if, at the same time, the interlocking of interests is actively maintained, then the law offers, as it were, a recall effect. After it is signed, a project becomes weightier, like a little sailboat whose hull has been ballasted with some heavy metal. It can still be overturned, but one would have to work *a little harder* to prevent it from righting itself, from returning to its former position. In the area of technologies, you cannot ask for more. Nothing is very solid in this area; nothing offers much resistance.* But by accumulating little solidities, little durabilities, little resistances, the project ends up gradu-

*See W. Bijker and J. Law, eds., *Shaping Technology—Building Society: Studies in Sociotechnical Change* (Cambridge, Mass.: MIT Press, 1992), especially the introduction.

ally becoming somewhat more real. Aramis is still on paper, but the paper of the plans has been supplemented by that of the patents—a plan protected by law—and now by that of the signed agreements. The courts are behind Aramis from this point on, not to say what it has to become, but to make it more difficult for those who have committed themselves to it to change their minds, to hold onto their money, or to back out of the project if the going gets rough. Yet "make it more difficult" does not mean "make it impossible." Woe betide those who trust the law alone to shelter their projects from random hazards.

"Apparently," Norbert said, "they all seem to be clinging to the idea that in guided transportation a mobile system that doesn't have a site isn't worth much. So we have to go to Orly, just as Aramis did. What is an 'exclusive guideway,' anyway?"

I was beginning to worry about this long line of "actors." If we had to go into the suburbs, we were going to lose a huge amount of time, or else spend a fortune on taxis.

"Keep the receipts for your expenses, my boy. I'll reimburse you; the client is paying. Ah, if only you were an ethnologist, you could stay in your village and draw nice neat maps. Whereas we sociologists have to drag ourselves around everywhere. Our terrains aren't territories. They have weird borders. They're networks, rhizomes."

"What?"

"Rhizomes, Deleuze and Guattari, a thousand plateaus."

The word "rhizome" wasn't in the dictionary. I learned later that *Mille Plateaux* (A Thousand Plateaus) was the name of a book and not one of Captain Haddock's swear words.

[INTERVIEW EXCERPTS]

M. Henne, head of the bureau of technological studies of Aéroport de Paris, speaking in one of a group of prefabricated buildings on the far side of the Orly runways. Through the window, the darker form of a test loop was identifiable in an otherwise empty lot.

"Yes, that's the place, Aramis was there. Why did we get interested? For me, Aramis isn't a PRT type of transportation, it's a short-distance transportation system. 'Hectometric,' we call it. When the Charles de Gaulle Airport at Roissy was being designed, we looked everywhere for a transportation system suitable for short distances. We even set up a company for moving sidewalks. We looked at everything.

"Then we got a new director. We gave up the project, we sold off our research-and-development markets and liquidated our subsidiary.

"But since we had been the driving force behind short-distance transportation, we had become experts of sorts. That's where the Aramis product comes in."

"For you, Aramis belongs to hectometric transportation?"

"Oh, yes, completely. Anyway, it's in the same sphere. Matra sought us out. We had a meeting in Lagardère's office. You have to admit he does a fine job of selling his products. He told us: 'That type of transportation is what we're good at. In Lille, it's working like a charm; moving sidewalks will never make it. You've invested a lot in short-distance transportation. You can't pull out now.'

"So we put a million into the deal. We paid in kind: the site, first of all, and the logistical support. I was head of the department of general studies for Aéroport de Paris. I was the one who followed the thing. They convinced us, but *we weren't a vector.* I myself was supportive, but it wasn't an immediate investment, only something *to check out, second-hand.*

"You know, I never believed in Aramis outside of airports. As soon as you enter the Paris region, you have to run the gauntlet of dozens of administrative offices. You never get anywhere. Anyway, after the oil crisis, all the short-distance systems fizzled out like balloons. Before, they were springing up all over. Remember the FNAC? There were a hundred such systems at the time. Then, in 1975, everything ground to a halt. People were looking for simplicity. They came up with Orly Rail, Roissy Rail. 'They'll be satisfied with that.'

"And I have to tell you something else. From that point on, all the contractors were going through a real panic over security issues, I mean passenger security. With Aramis, the issue came up over and over. I heard it dozens of times. What do you do in a car with some guy who looks suspicious? The demand just caved in.

"Before 1975, there was a period of innovation—new cities, all sorts of wild gimmicks. After 1975, it was all over; security was the only thing that counted . . . Besides, obviously, Aramis had to be done. Was it doable? I really don't know.

"But you know, I still tell myself that if somebody came up with the idea of the automobile today and had to go before a safety commission and explain, I don't know, let's say, how to get started on a hill . . . ! Just think how complicated it is: shifting gears, using the hand brake, and so on. He wouldn't stand a chance! He'd be told: 'It can't be done.' Well, everybody knows how to start on a hill! It's the same with Aramis. We hadn't gotten all the kinks out, but yes, I think it was doable." [no. 23]

To translate is to betray: ambiguity is part of translation.

For Aéroport de Paris, Aramis is not going to replace automobiles, remodel our cities, or protect our children's future. It is "in the same sphere as," in the neighborhood of, a much more modest means of transportation that involves a few hundred meters in airports, parking lots, or the FNAC on the rue de Rennes. It is limited to closed sites where one can innovate without competing with the heavyweights, with subways or trains. Moreover, for Aéroport de Paris, Aramis is not irresistible. Their people do not commit themselves one hundred percent; for them, it is not a matter of life and death. They took up second-line positions, to see what would happen. Is this the same Aramis as Matra's, or Bardet's? No, and this is precisely how a project can hope to come into existence. There is no such thing as the essence of a project. Only finished products have an essence. For technology, too, "existence precedes essence." If all the actors had to agree unambiguously on the definition of what was to be done, then the probability of carrying out a project would be very slight indeed, for reality remains polymorphous for a very long time, especially when a principle of transportation is involved. It is only *at the end of the road,* and locally, that the project will acquire its essence and that all the interviewees will define it in the same terms, differing only in viewpoint. In the beginning, on the contrary, it is appropriate for different groups with divergent interests to conspire with a certain amount of vagueness on a project that they take to be a common one, a project that then constitutes a good *"agency of translation,"** a good swap shop for goals. "Yes, for a million francs in kind, that's not bad. Why not go along? After all, it can't hurt." That's what

*See M. Callon, "On Interests and Their Transformation: Enrolment and Counter-Enrolment," *Social Studies of Science* 12, no. 4, (1982): 615–626; and M. Callon, J. Law, et al., eds., *Mapping the Dynamics of Science and Technology* (London: Macmillan, 1986).

Aéroport de Paris said. If you map out all the interests involved in a project, the vague or even reticent interests of those who are pursuing some other Aramis have to be counted as well. They are allies. Obviously, such allies are neither very convinced nor very convincing. As M. Henne says, they are not "vectors," and so they can drop the ball when things go badly. But if you had to have only associates who would stand up under any test, you would never stand up under any test.

"A concept, an innovation, patents, public authorities, an industrialist, an on-site installation at Orly—our Aramis is taking on consistency," Norbert exclaimed enthusiastically.

"They say they didn't have a user."

"No, they had an operator."

"What's an operator?"

"I suppose it's an operating agency, a business that really transports people, that's comfortable with mass transportation and can guarantee that the thing isn't dangerous. In France, there aren't many; it's got to be the SNCF or the RATP."

[INTERVIEW EXCERPTS]

At SOFRETU, a research arm of the RATP, the director is speaking.

"They needed an operator. The RATP said, 'Why not?' It was translated by financial participation.

"Obviously, the way cars connected made the Rail Division's hair stand on end, you can imagine—we nearly had a collision! But finally, for the operator, it was seductive. Operators don't like branchings. For operators, lines are ideal: you go from one end to the other and back, pendulum fashion. But you have to have branches to service the suburbs. Aramis solved all that. For us it was a terrific idea, since it solved the problem of fine-tuning the adjustment of supply to demand.

"It was really an innovation at the system level; it didn't have to do with the thing itself, with components. Besides, Matra had said, 'We'll take components that already exist on the market.' The invention was the operating system; the idea was that the technology would follow. It was really the passengers who were targeted, passenger comfort, and also the operator." [no. 17, p. 6]

"So things are shaping up pretty well," said Norbert. "We know we don't have to go back before the 1970s, since everybody all over the world was making PRTs at that time. We know the chain of interests that connects DATAR, Bardet, INRETS, and Aramis. We can even reconstitute—and this is unusual—the little intellectual shift that gave birth to the innovation: it was the merger of transfer machines and public transportation, then the eighty-one-box matrix, then the crosswise division of the AT-2000. We understand why Matra took up the banner—it was attempting to diversify, it was getting into VAL, and it had a better base than Bardet. Furthermore, we have no trouble getting enthusiastic ourselves, even fifteen years later, over a radical innovation, since we suffer from its absence every single day in Paris, as we go about our investigation! We have no trouble understanding why the people at Aéroport de Paris got their feet wet and even contributed some land without having a whole lot of faith in the project. And to top it all off, we know why the RATP had to be involved."

"It's not always as neat and tidy as this?"

"No, no, most of the time the origins are too obscure. And you have to go to an enormous amount of trouble to imagine what could have been behind such crazy inventions. In our lab we often study incredible cases of technological pathology. Here it's just the opposite: we don't understand why the thing doesn't exist."

"Yet it really is a corpse, and we've actually been asked to determine the cause of death."

"Precisely, my dear Watson, but we already know that the fatal cause won't be found at the very beginning of the project any more than at its very end. In 1973, if you'd had five million francs, you would have put them up!"

"Maybe not," I replied cautiously. "I probably would have bought myself an apartment first."

"Listen to that! Real estate! That's why everything is going to pot in this country. And he's an engineer!"

2 IS ARAMIS FEASIBLE?

"Now we can go on to Phase Zero, since that's where they are. Where's your diagram? Okay! Here's where we are, in black and white" [see Figure 1 in Chapter 1].

"If I'm not mistaken," my mentor went on, "our engineers from Matra, Automatisme et Technique, and Aéroport de Paris have been out in the beet fields by the Orly runways since 1973, asking three Kantian questions: 'What can we know about Aramis? What can we hope to get out of it? What is it supposed to do?' Oh, they're great philosophers! Nothing exists as yet except beets and a principle of continuous kinematics. A rather speculative principle, but the time is right and the arguments are good, so it stirs up some interest. The last step in this progressive slippage of interest is to translate it into cold hard cash, sixteen million francs at the time.* The beet field still has to be turned into a transportation system, which is roughly equivalent to turning a pumpkin into a coach. Who are the mice and the fairies? Who are the actors?"

I had worked with the professor for a while, so I was no longer surprised to hear sociologists throw theatrical terms around.

*The equivalent of 64 million francs in 1988. DATAR (representing the interests, properly understood, of the French in need of improvements) contributed 3.4 million, Aéroport de Paris and Air France (representing a mitigated interest in short-distance transportation systems) invested 1 million, the DTT (representing the Transportation Ministry) chipped in 1 million, the Ile-de-France region (here is someone representing local users at last) contributed 0.5 million, the RATP (representing technology) 0.362 million, and Matra (committing the properly understood interests of its stockholders to diversification in all directions) invested the largest amount: 10 million francs.

"According to the notes," I said, "among the 'actors' we ought to interview is Cohen, who's a graduate of Supaéro; he's a satellite specialist hired by Matra as production head. His classmate Frèque was selected to keep tabs on VAL. What's VAL?"

"It's the automated metro in Lille. Everybody keeps bringing it up when we talk about Aramis; we're going to have to look into it, too. Especially because Matra built it. It was begun around the same time, and the same engineers worked on both projects."

I sighed when I thought of all the interviews we had to do. Funny—I'd chosen sociology thinking it would be less work than the Ecole's other internships, like man-machine interactions or signal analysis.

"Then there's Lamoureux, fresh out of the Ecole Polytechnique and Télécom—just like me, if I may say so—with no preconceived notions about the world of public transportation, again just like me. The RATP picked him to work on the project. Matra hired M. Guyot, one of Bardet's engineers—he's a specialist in machine transfers—to head its future transportation branch."

"What about Bardet?"

"Apparently he kept one foot in the project as technical adviser. Not one of those engineers had any background in transportation. As I understand it, they all said to themselves: 'If we can build satellites, we can surely build a subway.'"

"But that's an advantage, too. They were ready to reinvent the wheel if they had to. Now comes the tricky part, my friend. We need to move on from easy sociology to hard sociology. It's our turn to reinvent the wheel. And there's only one way to go about it: we have to dig into the documents."

[DOCUMENT: INRETS REPORT, APRIL 13, 1972]

Article 2: Composition of the Prototype

The prototype will consist of:

—a segment of track 800 to 1,000 meters long, a fixed station represented by a platform, and a movable station that will facilitate inexpensive simulations of various uses: a workshop, a control post, a reception building, and a parking lot;

—five full-scale cars: two for passenger use, three reserved for measuring instruments.

This prototype will illustrate one of the possible ways the Aramis system can work and will demonstrate its adaptability to various real-life circumstances.

All the original components of the prototype, both the material elements (hardware) and the working processes and programs (software), will be designed in such a way that they can be transposed to later installations intended for actual operation.

Commercial operation of the system should thus be possible as soon as the anticipated development work has been completed as specified in the initial agreement and its codicil, without further delay or additional development programs.

The present codicil also applies to the trials and experiments carried out with the prototype, in conformity with Appendix 1 of the initial agreement.

The initial agreement is a continuation of agreement number 71-01-136-00-21275-01, between the prime minister and the Société des Engins Matra.

[INTERVIEW EXCERPTS]

M. Lamoureux, the RATP engineer heading the project at the time, recalls with feeling:

"We had never been through anything like that. We'll never see anything like it again. In six months, we went from paper to a prototype. We did everything: the site, the track, the cars—we invented it all.

"We tried everything. We were a gang of kids, applying what we'd learned in school—it was fantastic. We'd work till three or four in the morning.

"We weren't just pencil pushers like they were later on. We were really into it. We had to solve all the problems as we went along.

"The track was a concrete-and-steel sandwich: as soon as it got damp, the bolts popped out like rockets and landed 15 meters away. When the committee members came by, we were terrified they'd be hit." [no. 4]

M. Cohen, Matra's project head at the time, speaking in his spacious, large-windowed office in Besançon:

"It's also a question of the times, you know. I have trouble imagining an industrialist today who'd say, 'We don't have a medium-distance ultrasound link-up? Okay, let's go for it—we'll invent one. There's no motor on the market? Never mind, we'll develop one.' And it was all like that. Today, everybody sticks to his own job. People don't take so many risks.

"You have to realize that in six months we did the entire feasibility model. It wasn't a prototype; in fact, it was a full-scale model." [no. 45]

M. Berger, former RATP engineer who was acquainted with the project at the time, responding to questions:

"So, Aramis at Orly—what is it, exactly?"

"It's a 1,200-meter track with a shunting station, with three little yellow cars—user-friendly computers—running around it in dry weather [see Photos 3–5].

"Each little car is equipped with an arm that lets it push on the left or right guide-rail, at branching points, so it can turn."

"Switching isn't done on the ground, then?"

"No, it's done on board. The big challenge with Aramis is that the cars are autonomous; they don't touch each other, yet they work together as if they were part of a train. They have nonmaterial couplings—nothing but calculations. So you can imagine how autonomous they are.

"Every car has to know who it is. It has to receive instructions about speed—'Here you can speed up; here you have to slow down.' It has to monitor itself constantly, but it also has to know what car it's following, so it has to be able to see, or at least feel, what's in front of it, the way a bat does. It also has to know what's coming along behind.

"To see at a distance, we chose a long-distance ultrasound sensor; for short distances, a rotating laser bundle reflecting onto two catadiopters. If the car in front is too close, according to the ultrasound sensor, and the car behind gets the message 'Form a train,' it has to approach the lead car without bumping into it.

"They're in constant dialogue, since the ultrasound sensor in front and the ultrasound responder in the rear are both active.

"Then, when they've joined up in a train, the car controls itself by means of the optical sensor so it will maintain a constant speed without jerky movements that would shake up the passengers. When it approaches the station it has been assigned to, at the precise moment of coupling it has to put its arm out to the right or the left—pow!—while uncoupling from its colleagues at the same time *[he makes the gesture by turning in his chair]*. The cars behind have to close ranks *[he rolls his chair closer to his desk]*. If the first car in the train is the one that turns, the next one becomes first in line and gets its orders. The central computer keeps track of the flow in terms of passenger demand.

"So you see, it's not that easy. Each car has to calculate its own speed and position; it has to know where it's going; it has to be able to be leader or follower; it has to know when to stop at a station, open its doors, and keep track of passenger destinations. And take off again. If it's on its own, it can go full speed ahead. As soon as it sees a car in front, it has to be ready to meet and link up. All this, of course, at 25 kilometers per hour, even in rain or snow, all day, all night, thousands of times, without breaking down."

"And besides, what you are describing involves only three cars. A real system would need hundreds of brains like this."

"Yes, and what's more, the whole system has to be failsafe."

"I find your characters one-dimensional. They seem flat. They're just ideas, words on paper. They need to be animated; you have to make them move, give them depth and consistency. More than anything, they have to be autonomous; that's the whole secret. And instead they're so rigid! They'd pass for puppets. Look at that one: he has no personality, he doesn't know where he is, or what time it is, or where he's supposed to go, or whom he's supposed to meet. You have to tell him everything: 'Go forward, go back, come closer, turn right, turn left, open the door, go ahead, watch out.' Your characters are just sacks of potatoes. Give them a little breathing space, a little autonomy. Make them cars with minds of their own. You don't know a thing about art: it's not enough just to treat characters as vehicles for your projects. It's not worth it, being the best in the world in robotics, automation, mechanics, computer science, if you don't take the trouble to breathe some life into your anemic paper figures. Good literature isn't made with noble sentiments, Gentlemen, and good transportation isn't made with ideas. either. It has to have a life of its own: that's your top priority."

"But what if they start moving around on their own, taking their lives in their own hands? Maybe they'll get ahead of us!"

"And what are you getting paid for, may I ask, if it's not to come up with a transportation system that has a life of its own and can get along without us? If we always have to keep after it and tell it everything, if we can't ever hand a system over to our clients, keys and all, so that they'll only have to take care of upkeep and maintenance, we'll never make a cent. We're doing business, you know, not writing novels; we're supposed to outfit Jacksonville, Taipei, O'Hare, Bordeaux . . . These characters have to live on their own, do you hear me? They have to. You figure it out."

[INTERVIEW EXCERPTS]

M. Lamoureux:

"We had as much fun as a barrel of monkeys. We'd hold a catadiopter out in front of a car, on a stick, and we'd say: 'Heel, Fido,' and the car would come right up to the catadiopter and stop on a dime. When we moved ahead, it would follow.

"We even did a public demonstration. It was on May 3, 1973—I'll always remember that date—with Lagardère and a bunch of journalists. Everything had gone smoothly up till then. Of course, the day of the demonstration—it always happens—the thing didn't work. We'd wanted to do it all just right, all automatically, no tricks; but nothing worked.

"Impossible to get the system going. Lagardère was hysterical. 'I never should have organized this,' he said. 'Now the technicians will have to lick the bureaucrats' boots.' It's true that up to then we'd been doing whatever we wanted. After that, we had to work with fixed objectives.

"Naturally, just as you'd expect, a few minutes later we found the problem: everything worked fine. We wanted to call the journalists back, but Lagardère wouldn't hear of it." [no. 4, p. 6]

Innovations have to interest people and things at the same time; that's really the challenge.

The general director is interested; he has no trouble getting journalists interested, with the prospect of a scoop and some tasty hors d'oeuvres. Everybody gathers in the beet field that has been transformed into a revolutionary system of transportation for the year 2000. Just the moment for a mishap to occur, as so often happens in projects, although not so often in fairy tales: in no time at all, the coach turns back into a pumpkin, hesitates, turns into a coach again, then a pumpkin, and when it finally turns back into a coach again it's too late. The Big Interests gathered round are always in a hurry; they're tired of waiting, they disappear . . . The hors d'oeuvres are still there; as for the transportation system, sorry . . .

The problem is, the innovator has to count on assemblages of things that often have the same uncertain nature as groups of people. To get Aramis past the paper phase into the prototype phase, you have to get a whole list of things interested *in the project:* a motor, an ultrasound sensor, assorted software, electric currents, concrete-and-steel sandwiches, switching arms. Some of these actors and actuators, are docile, loyal, disciplined old servants; they don't cause any trouble. "I say 'come,' and here they are; I say 'go,' and they leave." This is the case with electric-power supplies, buildings, and tracks, even if the bolts do have an unfortunate tendency to pop out. But other elements have to be recruited, seduced, modified, transformed, developed, brought on board. The same sort of involvement that has to be solicited from DATAR, RATP, Aéroport de Paris, and Matra now has to be solicited from motors, activators, doors, cabins, software, and sensors. They, too, have their conditions; they *allow* or *forbid* other alliances. They require; they constrain; they provide. For example, in creating the cab, we can't follow the usual patterns of mass transit: the chassis would be enormous, and too heavy. We have to go see the manufacturer from Matra's automotive branch. But those people have never built a car on rails, or a car without a driver. So we have to get them interested, start from scratch: we keep the door, and add an in-transit switching mechanism to the cab. The motors, sensors, chips, and of course the software—none of these things is available commercially. We have to tie them in with Aramis: that's right, recruit them, sign them up, bring them on board. I may as well say it: we have to *negotiate* with them.

Nothing says that an ordinary electric motor, for example, was predestined to be used by Aramis, just as nothing says that Aéroport de Paris was predestined to use Aramis. Of course, standard electric motors can activate wheels, but not always under the very special constraints of nonmaterial coupling. Of course, Aéroport de Paris needs small-scale means of transportation, but not necessarily those particular means. These two actors, human and nonhuman, both have to be pampered and adapted so they can be put together in the project. It's the same task of involvement in both cases. The ordinary rhythms of these two actors turn out to be uneven, variable, interrupted. But that's where the danger lies. If they're seduced, convinced, transformed, pushed too far off their customary tracks, they may also become traitors and deserters. The motor won't work any more, and another one will have to be developed for the purpose. As for Aéroport de Paris, it will "stall." Everybody will lose interest, will pull away from the project. "Say, Aéroport de Paris, are you still going along with the project?" "Oh no, not if you get too far away from short-distance transportation systems! There's where I bow out!" "But hey, won't a good old motor work well enough for the project?" "Not a chance! Just a few tenths of a millimeter off, and it stops working!" The effort to generate interest has failed. The human and nonhuman actors have once again become admirably disinterested.

The full difficulty of innovation becomes apparent when we recognize that it brings together, in one place, on a joint undertaking, a number of interested people, a good half of whom are prepared to jump ship, and an array of things, most of which are about to break down. These aren't two parallel series that could each be evaluated independently, but two mixed series: if the "onboard logical systems" fizzle out at the crucial moment, then the journalists won't see a thing, won't write any articles, won't interest consumers, and no money or support will get to the Orly site to allow the engineers to rethink the onboard logical systems. The human allies will scatter like a flock of sparrows and they'll go back to their old targets—consumers to their cars, the RATP to its subway, Matra to its space business, and Aéroport de Paris to its Orly Rail systems. As for the nonhuman resources, they'll all return to their old niches—the ultrasound sensors will go back to the labs, the classic motors will go back into classic electric cars, the doors and windows will be beautifully adapted to the automobiles they should have stuck with all along. So if you don't want the transportation system to turn back into a beet field, you have to add to the task of interesting humans the task of interesting and attaching nonhumans. To the *sociogram,* which charts human interests and translations, you have to add the *technogram,* which charts the interests and attachments of nonhumans.

"For my part," the motor declares, "I won't put up with nonmaterial coupling. Never, do you hear me? Never will I allow acceleration and deceleration to be regulated down to the millisecond!"

"Well, as for me," says the chip, "I bug the CEO and his journalists. As soon as they want to break me in, I break down and keep them from getting started. Ah! it's a beautiful sight, watching their faces fall, and poor Lamoureux in a rage . . ."

"That's pretty good," says the chassis. "Me, on the contrary, I let them move me around with one finger. I glide right over the tracks, since I'm so light, and I actually even let myself be bumped a bit."

"Oh, stop pretending you're an automobile!"

"Hah! A chassis like that, you really have to wonder what she's doing in guided transportation . . ."

"Leave her alone, she's a bootlicker!"

"She's right," says the optical sensor, "I help the car, too, and it's even thanks to me that the motor can be put to work."

"'Thanks to me'—listen to him! As soon as the laser angle is too obtuse, he loses his bearings! And he talks about putting me to work!"

"You're the one who's obtuse—you can't do anything but break down. At least I authorize linkups," says the central control panel, "and furthermore I'm Compatible."

"'I'm Compatible'! Well, I guess I'd rather hear that than be deaf," says the base computer. "But all the software had to be rewritten just for him."

I was horrified by the mixed metaphors; my boss loved them. He seemed to like watching the engineers think about mixing humans and things.

"It's a confusion of genres," I said, forgetting my place. "Chips don't talk any more than Chanticleer's hens do. People make them talk—we do, we're the real engineers. They're just puppets. Just ordinary things in our hands."

"Then you've never talked to puppeteers. Here, read this and you'll see that I'm not the one getting carried away with metaphors. Anyway, do you really know what 'metaphor' means? Transportation. Moving. The word *metaphoros,* my friend, is written on all the moving vans in Greece."

[DOCUMENT: MATRA REPORT BY M. LAMOUREUX, END OF PHASE 0, JUNE 1973; EMPHASIS ADDED]

In order to be able to *ensure* the basic functioning of the Aramis principle, an Aramis car must be able to:

—*follow* a speed profile provided while it is moving independently or as part of a train ("proceeding in a train");

—*take* the turnoff to its target station ("removal of a car from the train");

—stop in a station individually or as part of a train ("station stop");

—*connect* with a car that has just left a station ("linkup");

—approach a car within a train after an intervening car has been pulled out, in order to close the gap ("merger").* [p. 7]

Every car thus has to *supervise* itself: the subsystems include *controls* that go into "onboard *security software*," which has *responsibility* for detecting any abnormal functioning of the car and for *ordering* an "emergency stop." The emergency-stop order is transmitted by a security transmission to the neighboring cars and to the "ground security software," which *informs* the central calculator of the emergency stop.

Paralleling these controls at the level of the cars, the calculator *supervises* the normal progression of the selected program and can also transmit an emergency-stop order to the ground security software.

All the cars in a train receive *instructions* about their speed, which are forwarded by the central calculator. The "functional onboard software" transmits these instructions to the automatic guidance system:

—if the car is at the head of the train, the automatic guidance system *brings* the real speed of that car *into line with* the assigned speed;

*"Linkups are different from mergers; in mergers, both cars are actually moving at the same speed when the operation begins. There are fewer constraints on mergers, but the same equipment brings them about." [p. 7 note]

—otherwise, it *instructs* the car to follow the one ahead at a distance of thirty centimeters. The "automatic pilot" of the car behind *knows* the distance that separates it from the car ahead, because it has a short-distance sensor (optical).

The "functional onboard software" (FOS) of each car has in memory an *authorization* for linkup or merging that is sent to it by the central calculator. If merging is *authorized,* any detection of a car at less than forty meters entails a merger behind that car. Every car has a "long-distance sensor" (ultrasound) with a scope of one to forty meters that supplies information about distances for the piloting system. A "short-distance sensor" (optical) is used instead for distances of one meter or less. [p. 9]

Certain orders intimately tied to security are determined *"by majority vote."* The FOS carries out a majority vote on the basis of five consecutive receptions when it recognizes such an *order;* these are orders *commanding* the switching arms to operate or *prohibiting* mergers. [p. 21]

Men and things exchange properties and replace one another; this is what gives technological projects their full savor.

"Subordinate," "authorize," "supervise," "allow," "notify," "possess," "order," "vote," "be able": let's not jump too quickly to conclusions as to whether these terms are metaphorical, exaggerated, anthropomorphic, or technical. The people interested and the machines recruited don't just get together on a joint project so they can bring it from the paper stage to reality. Some of them still have to be *substituted* for others. Aramis, for example, can't be controlled by a driver as if it were a bus or a subway, because each car is individualized; in the initial stage of the project, each had only four seats. You can't even think of putting a driver—a union member to boot—in every car; you might as well go ahead and offer every Parisian a Rolls. So something has to take the place of a unionized driver. Will the choice be an automatic pilot, with its "functional onboard software," or a central computer with its omniscience and its omnipotence? In moving from humans to nonhumans, we do not move from social relations to cold technology. For *some* features of human drivers have to come along and stay on board, or else they have to come from the center. We won't keep the humans' physical presence, their caps, their uniforms, or their outspokenness; but we'll keep *some* of their knowledge, their abilities, their knowhow. Cold qualities? No, on the contrary, warm and controversial, like subordination and control, authorizations and orders. Because the automatic pilot is demanding as well—not about retirement and Social Security, but about distance sensors, orders, and counterorders, if we decide to put it on board; about transmission, road markers, information, and speed, if we set it up at the command center. When our engineers cross the qualities of drivers with the qualities of automatic pilots and central computers, they're embarking on the definition of a *character.* An autonomous being or an omniscient system? What minimum number of human qualities does that character have to bring along? What characteristics have to be delegated to it? What sensations does it have to be capable of experiencing? Yes, we're actually dealing with metaphysics, and the anthropomorphic expressions must be taken not figuratively but literally: it really is a matter of defining the human *(anthropos)* form *(morphos)* of a nonhuman, and deciding on the limits to its freedom.

[INTERVIEW EXCERPTS]

M. Berger, reading the above document aloud: "Ah, that's what I was looking for: 'Aramis is an automated transportation system that includes a maximum number of *onboard controls;* as a result, transmissions have been cut down *to a minimum*. The vehicles have been made completely automatic, and the calculator *addresses* all the vehicles present on a functional transmission segment' (p. 24). Yes, I have to admit that our discussions sometimes took on theological dimensions."

"There, Father, you can see in the ground controls a precise model of God's relations with his creatures."

"But isn't it rather impious, my son, to represent God as the constant repairer of his creatures' mistakes? The world you propose is hardly perfect, since, according to you, the Supreme Intelligence has

not only determined the laws according to which the world works but is also required to correct them constantly. For my part, I'd prefer a system more in conformity with that of Mr. Leibniz, one in which God's creatures would contain the complete recapitulation of all possible actions. It would suffice to enter all predicates in the software. So, for example, if the creature 'Julius Caesar' were opened up, an infinite intelligence could read everything he will necessarily do—from his birth and adoption to the Rubicon and the Ides of March. In the same way, by opening up the prototype creature, you could deduce all degrees of speed, all bridge crossings, and all station stops. The prototypes, like true monads, would have no doors or windows."

"A serious drawback for the passengers, don't you agree, Father?"

"There is the problem of the passengers, of course [he chuckles monkishly] . . . But then, wouldn't a project like that be more worthy of God's greatness and perfection than your Malebranchean universe, which achieves harmony only through constant repair? Whereas I would achieve it through perfect calculations, and all the prototypes would go their own ways because of the preestablished harmony in their software; they wouldn't have to see or know each other. Don't you agree, my son, that this world would correspond more closely to the picture that piety should draw of God?"

"Of God, no doubt, Father, but how about Matra? Even an inertial platform couldn't keep its fixed point without being reinitialized from time to time. You're asking too much of human beings."

"And you, my son, are not asking enough of God."

"But what do you do about freedom, Father? Why not allow the vehicles enough knowledge to take care of harmonizing the laws of the universe—fixed by God—with the little adjustments that human imperfection and sin have put in Matra, in the chips as well as in the tiniest little fleas? Why not open up our monads? Let's give Aramis more autonomy, as befits a divine creature, after all; for won't God's work be judged all the more beautiful to the extent that His creatures are more free? Instead of making them automatons, as you do, I'd make them living creatures. They'll know how to repair themselves, and they'll get their bearings from one another. Instead of communicating abstractly with their Creator, as you propose, they'll find a new harmony owing to their freedom. Yes, Father, they'll be connected by a vinculum substantiale. *Nothing material will link them together to keep them on the right path. They'll have to make independent decisions, check themselves, connect and disconnect, in conformity with the laws*

of the world system to be sure, but freely, without touching each other and without being the slaves of any automated mechanisms. Oh, what a beautiful construct! How much worthier of inspiring piety in the atheist's hardened heart than the fatum mahometanum, *the predestined world you depict!"*

"You're getting pretty hot under the collar, my son. I detect very little piety in the culpable passion that makes you want to create living creatures yourself."

"Forgive me, Father, I did get carried away, but the questions of freedom and predestination are ones I care deeply about."

"Where is this chapter on the preaffectation of stations?"

"Oh—sorry, Lamoureux, I was thinking about grace."

"Grace?"

"I mean that you can't let passengers decide for themselves where they're going. The central system has to decide for them."

"There's still a problem, though," I said as we left Berger's office. "Obviously, the Orly track is only 1,200 meters long, and the bolts do come loose, but still, the three vehicles really exist, and they get their bearings from each other by means of optical sensors, and the couplings are calculated. Why doesn't the story end at Orly," I asked, mystified, "since Aramis is feasible? Why do we have to raise the question of its existence fourteen years after the fact, if it was already completed in 1973? What's more, some of our interviewees didn't even mention Orly. For them, the story started in 1984!"

"That part doesn't bother me too much. The body of a technological object is made up of envelopes, layers, successive strips. A project never stops becoming real. It's normal for people just coming on board to be ignorant or scornful of the past. What bothers me the most, you're right, is the impression that we're losing in terms of reality. Things happened at Orly that no longer look feasible fifteen years later. The old guys like Lamoureux and Cohen were ahead of the technicians like Frèque and Parlat, who came later. That's much less common."

[INTERVIEW EXCERPTS]

M. Lamoureux, responding to questions in his office overlooking the Seine:

"At the end of 1987, when Aramis was rolling, it seemed very hard to get three cars to move together for more than a few seconds. There was a pumping effect, and they bumped into each other; the train tended to come apart. Do you remember seeing three cars working at Orly in 1973? Weren't there only two?"

"No, the three were there. Let's see . . . Now that you raise the question, I'm suddenly having doubts, but I still see us with the three . . . Wait a minute, I'm going to ask *[he phones a former Matra colleague]* . . . She remembers the three cars quite well; she says there must even be a videotape . . ."

"How do you account for the fact that things could be done in 1973 that still seemed borderline possibilities last year?"

"Listen, I want to tell you something. I'd rather you didn't write this down, but we were doing better in 1973 than in 1987, fourteen years later, and with primitive electronics." [no. 4]

Berger:

"There's no comparison between Orly and what had to be done afterward, because Orly was not *failsafe*. If you take away the guarantees, you can make any idea work, as long as you're always on hand to make repairs and start it up again. In fact, when there are ten engineers for three cars, it's not really what you'd call automation! But 'failsafe' means being in regular use, like the Paris metro, day and night, with passengers who break everything, and just regular maintenance. To compare a prototype like the one at Orly with a real transportation system is meaningless." [no. 14]

In M. Cohen's office:

"People often say that Orly wasn't representative because Aramis wasn't failsafe."

"That doesn't mean anything. In any case, it depends on what you mean by failsafe. If it means *intrinsically* failsafe, then there's no question: Aramis can't work at all as an intrinsically failsafe system; it couldn't fifteen years ago, it can't now, it can't twenty years from now. If it's probabilistic safety you're talking about, then yes, we were failsafe at Orly. We showed that it worked." [no. 45]

In M. Etienne's office at Matra:

"Weren't you already doing all that in 1973—operating the cars in train formation, merging them, pulling one out, and for more than just a few seconds?"

"At Orly, we did a functional demonstration, but not under failsafe conditions. I myself wasn't aware that it wasn't failsafe in 1974. I want to be very frank about this; I learned it later. Had the engineers concealed the fact that it wasn't failsafe, or had I just not seen it? I don't know. In any case, I wasn't aware of it. Matra had already spent 10 million francs on it. Everybody was very proud; we had shown that Aramis was feasible." [no. 21]

M. Parlat:

"Orly didn't prove a thing. But it's true, people thought that Orly proved Aramis' feasibility once and for all, that there would be no need to go back over that ground." [no. 2]

The reality, feasibility, and representativeness of a project are progressive concepts, but they are also controversial; that's why it's so hard to get a clear idea about the technologies involved.

"Aramis is feasible; Orly proves it." "Orly doesn't prove a thing. Aramis isn't feasible." Depending on whom you talk to, Orly gets lost in the mists of time and takes on the position of a simple idea, as brilliant but as unreal as Bardet's eighty-one-box matrix, or else it provides such thorough proof of Aramis' feasibility that all that's left to do is find a few billion francs in order to carry millions of passengers in the system—assuming a few minor improvements can be made.

No one sees a project through from beginning to end. So the tasks have to be divided up. Let's suppose Bardet says, "The bulk of the work is behind us. I've got a patent on Aramis." Statements like this would make Matra laugh, since both Gayot and Cohen start by abandoning Matra's solutions. They start from scratch, and in six months they do the bulk of their work: they make a life-size model. Let's suppose that they declare in turn, "Aramis exists; the hardest part is behind us. All we need to do now is settle a few minor issues and fine-tune the whole thing." We can hear the guffaws in the RATP's railway division: "Fine-tuning? But Aramis doesn't exist, because it isn't failsafe. What you have there is no more than an idea that is possibly not unworkable." And indeed, if the project continues, they have to change the motor, multiply the redundancies, redo all the software, redesign all the chips. But the laughter may keep on coming. Let's imagine that all the aforementioned engineers are congratulating themselves on their success and breaking out the champagne to toast Aramis. Others, at the RATP, will be laughing up their sleeves: "A transportation system isn't just a moving object, no matter how brilliant; it's an infrastructure and an operating system. They're toasting Aramis, but they don't have a single passenger on board! A prototype doesn't count. What counts is a system for production and implementation, an assembly line; and there, more often than not, you have to start all over again." Still others, instead of laughing, will get upset and pound their fists on the table: "A transportation system exists only when it begins to make a profit and when it has lasted without a major breakdown for at least two years."

The frontier between "the bulk of the work" and "fine-tuning the details" remains in flux for a long time; its position is the object of intense negotiation. To simplify its task, every group tends to think that its own role is most important, and that the next group in the chain just needs to concern itself with the *technical details,* or to apply the principles that the first group has defined. Moreover, this way of looking at things is integrated into project management: by going from what is less real to what is more real, you often divide up projects into so-called phases: the conceptual phase, the feasibility phase, the scale-model phase, the full-system site study phase, the commercial-demonstration phase, the acceptance phase, the phases of qualification, manufacturing, and homologation.

If Aramis fit into this grid, there would at least be a regular progression. Unfortunately, not only are the phases ill-defined, but they may not come in order at all. People who are studying a project may indeed disagree about the sequence. Does Orly prove that Aramis is feasible? Yes, if you believe those who define Aramis as a moving object endowed with original properties. No, if you believe those who define it as a transportation system that can be set up in a specific place that meets certain use constraints. You can even argue over whether *more* was being done in 1973 than in 1987. So it is possible to imagine that you lose in terms of the project's degree of reality over time. Consensus about the length, importance, and order of phases is not the general case. It is a special case—that of projects that work well. With difficult projects, it is impossible to rely on phases and their neat arrangements, since, depending on the informant and the period, the project may shift from idea to reality or from reality to idea . . . This is something Plato didn't anticipate. Depending on events, the same project goes back into the heaven of ideas or takes on more and more down-to-earth reality. Aramis (since it failed, as we know) has become an idea again—a brilliant one—after nearly becoming a means of transportation in the region south of Paris. There is obviously no way to contrast the world of technology, which is real and cold, efficient and profitable, with the world of the imagination, which is unreal and hot, fantastic and free, since the engineers, manufacturers, and operators all squabble over the definitions of degrees of reality, feasibility, efficiency, and profitability of projects.

[INTERVIEW EXCERPTS]

At RATP headquarters, M. Lamoureux is still responding to questions in his office overlooking the barges:

"Is Aramis feasible? 'Presumably,' some reply, 'since it worked at Orly.' 'Not at all,' say the others, 'since it isn't failsafe.' Forgive my ignorance, but I don't understand the way you're using the term 'failsafe.' Why does it make such a difference?"

"I can explain it to you very simply. We undertook a thoroughgoing analysis of the causes of breakdown in the standard metro system. That had never been done before in a serious way. At the time, that information was completely confidential.

"Well, when we did our tallies, we found that system breakdowns were very rare: 3 percent. I still have the figures. Next comes trouble with the passengers; that accounts for roughly 20 percent. All the rest of the breakdowns are caused by interactions between passengers and the system; that accounts for 77 percent, and of those breakdowns half are due to problems with the doors.

"This is for the part of the Paris metro system that runs on automatic pilot, but there's still a driver who takes over when something goes wrong. Okay, now look at what happens if you go completely automated. Obviously, it's out of the question for an automated system to do less well than the metro. On VAL, with smaller trains, availability has to be increased by a factor of 100 if there are to be no more breakdowns. That's hard, but it's doable.

"If we were to go on to Aramis, which at the time carried four people per car, availability would have to be increased by a factor of 5,000! In addition, there would be a lot of new functions to take care of: the linkups, as well as the motor, which was new.

"And even though microprocessors were becoming more common, they weren't reliable. I said so at the time: we don't know how to do it, we're running up against technological limitations. We could have built in redundancies; but if you do that, the costs get out of hand . . .

"The rest of the story is a series of steps backward by the decisionmakers, who were hiding behind economic studies. In the development of a system, especially one like Aramis, there's a huge component of speculation. Economic studies have their place, but *you shouldn't do too many of them* at the beginning. To start with, all you should have to do is stay within some limits.

"In any event, Giraudet went around the table and I said that Aramis wasn't generalizable in the absence of a technological mini-revolution. It wasn't worth doing economic studies, since *the thing was technologically unworkable.*" [no. 4, pp. 9–10]

M. Berger:

"You have to understand intrinsic security. That's the underlying philosophy of the SNCF and the RATP. What it means is that, as soon as there is any sort of problem, the system goes into its most stable configuration. It shuts down. Broadly speaking, if you see the subway trains running, it's because they're authorized *not to stop!* That's all there is to it: they're always in a status of reprieve.

"As soon as this authorization stops coming through, the trains stop running. The emergency brake comes on, and everything shuts down. So everything has to be designed from A to Z—everything, the signals, the electric cables, the electronic circuits—with every possible type of breakdown in mind."

"That's why they call it *intrinsic:* it's built into the materials themselves. For example, the relays are specially designed so they'll never freeze up in a contact situation. If there's a problem, they drop back into the low position; their own weight pulls them down and they disconnect. The power of gravity is one thing you can always count on. That's the basic philosophy.

"Even if you're dealing with electronics, you have to be *absolutely* sure, not just *relatively* sure, that all possible breakdowns have been identified, in the hardware and software both. And here's where it's tricky, because we don't know how to check software—it's too complicated.

"We've been checking switches and electric signals for a hundred years. There are procedures, committees, an incredibly precise methodology. Even so, every year we still find mistakes. So you can imagine, in software that has thousands of instructions and that is cobbled together as fast as possible by consultants . . .!" [no. 14]

M. Cohen, at Matra headquarters in Besançon.

"There was no hope for Aramis if you had to bring in the principles of intrinsic security. Not a chance."

"I thought that was the basic philosophy in transportation."

"Not at all. An airplane doesn't have intrinsic security. Just imagine what would happen in an airplane if everything came to a halt whenever there was a minor incident! Well, people take planes, they accept the risk; this is *probabilistic* security. It was the same with Aramis.

"And let me tell you something: the RATP was ready to take the risk. They were much more open than people have said; they were ready to change their philosophy.

"I can write the equation for you *[he takes out a piece of paper and writes]:*

Probabilistic security = Aramis is possible;

Intrinsic security = Aramis is impossible.

"It's as simple as that. But I've finally concluded that in transportation, the only philosophy that allows a decisionmaker to make a decision is intrinsic security. Not for technological reasons. When you say 'intrinsic,' it means that, if there's an accident, people can say: 'Everyone involved did everything humanly possible to provide a response for all the possible breakdowns they were able to imagine.' This way the decisionmakers are covered. They can't be blamed for anything.

"In probability theory, you say simply: 'If event *x* happens, and if event *y* then occurs, there is a risk of *z* in 1,000 of a fatal accident.' And this is accepted, because the probability is slight. This approach is unacceptable for a decisionmaker in the field of public transportation."* [no. 45, p. 2]

M. Chalvon, managing director of Alsthom at the time of the Aramis project:

"Of course, I had my technical services study Aramis at the beginning. You always have to check whether your competitors aren't about to get ahead of you. I remember the technical report.

*The small number of deaths per year that can be blamed on public transportation all make headlines. The 12,000 annual fatalities in France that can be blamed on automobiles—a transportation system whose security is probabilistic or even random!—don't make headlines. Hence the obsession with security characteristic of guided transportation systems.

"The idea was that it was seductive, but impossible given the constraints that characterize mass transportation. In some other civilization perhaps, but not at the end of this particular century.

"Matra didn't realize at the time what was involved in the world of public transportation. They had their apprenticeship with VAL. At Alsthom, we've been involved in guided-transportation systems for a long time. There are far more constraints on them than on satellites.

"We're used to factoring in the sort of constraints that are brought to bear on a system by millions of fed-up passengers. Not only do you have to protect yourself against impulsive acts, but the vandalism is unbelievable. You have to be able to defend yourself against it.

"Cars belong to individuals; everyone looks out for them. But Aramis would have been collective property. The first time anything went wrong, people would have blown the whole thing up." [no. 46]

We are never as numerous as we think; this is precisely what makes technological projects so difficult.

Not only do the actors vary in size, so that they may represent fewer allies than they claim to stand for, but they may also bring into play far more actors than anticipated. If there are fewer of them, the project loses reality, since its reality stems from the set of robust ties that can be established among its actors; if there are too many of them, the project may well be swamped by the erratic intentions of multiple actors who are pursuing their own goals. For a project to materialize, it must at once recruit new allies and at the same time make sure that their recruitment is assured. Unfortunately, discipline isn't the strong point of Parisians, programmers, decisionmakers, or chips. There are breakdowns, there are damages, there are impulsive acts, there are dead bodies; there are trials, decisionmakers on the stand, articles in the newspapers. "We hadn't anticipated this," say the operators. "You should have," replies the angry crowd. "We hadn't taken all these problems into account," say the people behind Aramis. "You have no choice but to take them into account," say the people in charge of security, implementation, siting.

A transportation system is no better than its smallest link. If it is at the mercy of a vandal or a programmer or a parasitic spark, it isn't a transportation system—it's an idea for a transportation system. To the task of generating interest, which linked up a crowd with a project, is now added the task of *protection;* this consists in rendering harmless the behavior of a different crowd, made up of intruders, bugs, troublemakers who've shown up uninvited. The system has to be made idiot-proof. This applies to relations among human beings—you have to prevent passengers from coming to blows. It also applies to relations among humans and nonhumans—you have to prevent people from getting caught in doors, and from being able to jam the doors. It applies as well to relations among nonhumans—you have to prevent bugs in the chips from continually setting off the emergency brake. It applies to a completed transportation system that is actually working full scale, and it applies to a project for developing a transportation system that is designed to work full scale. The difference between the two is precisely the *taking into account* of an infinite number of unanticipated details that have to be mastered or done away with one by one. This is the beginning of a new set of negotiations whose success or failure will make it possible to modify the relative size of the project over time.

Human error is everywhere; so is diabolical wickedness; and imbecility is common. As for software programs, they go right on making mistakes without anyone's being able to identify the bugs that have infested them. Here is the difference between a project that is not very innovative and one that is highly innovative. A project is called innovative if the number of actors that have to be taken into account is not a given from the outset. If that number is known in advance, in contrast, the project can follow quite orderly, hierarchical phases; it can go from office to office, and every office will add the concerns of the actors for which it is responsible. As you proceed along the corridor, the size or degree of reality grows by regular increments. Research projects, on the other hand, do not have such an elegant order: the crowds that were thought to be behind the project disappear without a word; or, conversely, unexpected allies turn up and demand to be taken into account. It's like a reception where the invited guests have failed to show; in their place, a bunch of unruly louts turn up and ruin everything. In this sense, Aramis is unquestionably a research project.

The innovator's work is very complicated. Not only does she have to fight on those two fronts, dealing with supports that are removed and parasites that are added; not only does she have to weave humans and nonhumans together by imposing the politest possible behavior on both; not only does she have to attach nonhumans together; but she also has to know who, among the engineers, executives, and manufacturers speaks for the good actors that need to be taken into account. Should the managing director order a market study—which would speak in the name of consumers—when his technical department is declaring that the project is not technologically feasible without a revolution in microprocessors? Whom should he believe? His safety division, which is claiming that you can't make a transportation system without intrinsic security, or his commercial developers, for whom vandalism rules out sophisticated systems? Of course you have to "take into account" all the elements, as people say naïvely, but only the not very innovative projects know in advance *which accountant* to believe and *which accounting system* to choose. We use the term "innovative" precisely for a project that requires choosing the right accountant and the right accounting method, in order to decide which actors are important and which ones are dangerous. By this measure, yes, Aramis is decidedly an innovative project.

"But we have the same problem," I said, rather discouraged, after reading his commentary in the train that was taking us to Lille to visit the VAL system. "Who are we supposed to take seriously? Lamoureux places a lot of stress on the failure of the demonstration with Lagardère at the Orly site. The others tell us that that had nothing to do with it. Not until the forty-fifth interview, the one with Cohen, do we hear that the connection between probabilistic security and Aramis is an absolutely crucial link that explains the whole project. Other people are telling us that 'in any event, in the Paris region, you can never introduce radical innovations in public transportation.' 'In any event,' others say, 'it's not a transportation problem; you can never innovate across the board, and Aramis involves innovations across the board, in everything: components, functions, manufacturing, applications. And the other guy, Chalvon, is telling us that you can't do anything about vandalism. It's really kind of discouraging. How can you claim to explain an innovation by taking into account the determining elements if there are as many lists and hierarchies and ways of accounting for these elements as there are interviewees?"

"We are de-sorcerers, my friend, not accusers. We deal in white magic, not black. Each of these accusers is pointing a finger and blaming someone or something. Our job is to take all these sorcerers into account. If the project had succeeded, they'd have more to agree about. Look at VAL: everybody is rushing to shore up its success, and everybody agrees about the reasons for it, even if they divide up the credit differently. It's only because the project failed that blame is flying in all directions. No, these discrepancies don't bother me. We aren't going to settle our accounts in their place. What we have to understand is why the fate of the project didn't allow them to come to an agreement even on the same accounting system. Why isn't there any object which they can see eye-to-eye on?"

"Wait a minute, do you mean you're not going to tell us after all who killed Aramis? You're not going to find the guilty party? But that was the deal! You're cheating! Columbo always figures it out; he gets all the suspects together and points his accusing finger at the one who's guilty. He holds up his end of the bargain."

"Fine, but in the first place Columbo is a fictional creation, and in the second place he deals with human corpses and their murderers, whereas our job is to look for dismemberers of assemblages of humans and nonhumans. Nobody has ever done that before. Except in a couple of nineteenth-century masterpieces—which you have to read, by the way. One is by Mary Shelley . . ."

"Mary Shelley? Isn't she the one whose imagination ran away with her? The one who succumbed to her own fictional creatures?"

"Yes, she was the daughter of the first great feminist, the wife of the poet, the mother of Frankenstein."

"And the other one?"

"Butler's treatise. In *Erewhon* they got rid of all the machines because, like all good Darwinians, they were afraid that machines would take over. Erewhon, my friend, is Paris and its intelligentsia. They've wiped out all technologies, large and small; intellectuals never give them a thought. And it works out perfectly: our finest minds, our most exquisite souls really do live in Erewhon."

We arrived at the train station in Lille. All day we had the pleasure of admiring the architecture, the slow, steady glide of driverless trains, the remote guidance from the command post [see Photo 7].

[INTERVIEW EXCERPTS]

"I'm giving you my own personal version . . ." [no. 32]

"I'll try to reconstruct what I thought at the time . . ." [no. 29]

"It's hard to be objective. Here's what I could see from my window . . ." [no. 20]

"I'm not speaking from Mars; I hope you don't expect anything more from me. This is just my own point of view, that of a minor local representative . . ." [no. 44]

"What I have to say is very subjective. I'll just tell you everything all jumbled together, and you'll have to sort it out . . ." [no. 12]

"With hindsight, that's what I'd say, but the president doesn't see things the same way." [no. 42]

"I'm not here to tell you the truth, just how we felt about things at the time." [no. 37]

"You know, in a case like this, you just have to speak for yourself, since our own higher-ups don't agree. Bill and I are in the same department; we see things more or less the same way. You have to get right down to individual opinions." [no. 30]

About technological *projects,* one can only be subjective. Only those projects that turn into objects, institutions, allow for objectivity.

Is the Aramis story already over, or has it not yet begun? The interviewees can't settle this issue. After all, we can step aboard VAL, in Lille; we can't step aboard Aramis, in Paris. Hundreds of thousands of Lille residents head for VAL every morning: they go to its stations, follow its signs, learn how to pay, wait on the platforms in front of closed doors, sense the train gliding past the glass windows in a white blur; they see doors opening, climb into the train, listen to announcements made by a synthetic voice, and look at the doors that close before they are carried off into dark tunnels. Actually, that's not true; they see nothing, feel nothing, hear nothing. Only tourists are still surprised to see a simple placard in place of a driver and are a little shaken up by the station stops required by the intrinsic security system. Lille residents have come to take VAL so much for granted that they no longer think about it, no longer mention it when they want to go from one place in the region to another. VAL goes without saying. "M. Ferbeck? M. Ficheur?" No, those names mean nothing to them. Yet those are the names of VAL's creators, its developers, its stage managers! "No, really, we don't know a thing about them." More fortunate than Parisians, Lille residents don't even have to recall the existence of humans on strike days, since the automatic pilots aren't unionized. Only on those rare occasions when the system is down do Lille residents remember that "they" exist and that "they" are about to come and fix things.

No one takes Aramis in the thirteenth arrondissement in Paris, or the fourteenth or the fifteenth, or in Nice, or in Montpellier, or even at Orly in the middle of a beet field. Few people think about Aramis—not because it has become so obvious that it no longer counts, but because it has become so inconspicuous that it no longer counts. In 1988 Aramis exists as a thorn in the side of the RATP, Matra, the Transportation Ministry, and the Budget Office, which is still wondering how the adventure ended up costing nearly half a billion francs. Aramis becomes a textbook case for the Ecole des Mines. It tugs painfully at the memory of some thirty engineers who have given it the best years of their lives. Dispersed among thousands of gestures, myriad reflexes, and immense know-how on the part of Matra and RATP engineers, it survives, but in a state that leaves it unrecognizable. People say, "Let's take the Aramis case." They don't say, "Let's take Aramis to the boulevard Victor."

VAL, for the people of Lille, marks one extreme of reality: it has become invisible by virtue of its existence. Aramis, for Parisians, marks the other extreme: it has become invisible by virtue of its nonexistence. The VAL *project,* full of sound and fury, arguments and battles, has become the VAL *object,* the institution, a means of transportation, so reliable, silent, and automatic that Lille residents are unaware of it. The Aramis project, full of sound and fury, arguments and battles, has remained a project, and becomes more and more so; soon it will be nothing more than a painful memory in the history of guided transportation.

The VAL object gathers to itself so many elements that it ends up existing *independently* of our opinion of it. Of course, the descriptions by the Lille residents, by Notebart (who is its father), by the supervisors in the control room, by Ferbeck (who is also its father), by Lagardère (who is also its father), by the RATP (which claims to be its father), by Alsthom (which also claims to be its father)—all these descriptions are going to vary, and especially on the question of paternity! There are as many points of view as there are heads. But these points of view are all focused on a *common object,* as if, while walking around a statue, each person were offering a different description that was nevertheless compatible with the others. Except for the point of view, the description is the *same.* VAL, because it exists, unifies points of view. It transforms people's opinions of it into "simple" points of view about an object that remains independent of them. With Aramis there is nothing of the sort. Since it does not exist, it cannot unify points of view. There are as many possible Aramises as there are minds and points of view. Matra's point of view and the RATP's and the ministry's are irreconcilable; within the RATP, within Matra, within the ministry, individual opinions are equally irreconcilable; the same Matra engineer, from one interview to another, or the same ministry official, between the beginning of an interview and the end, will also have irreconcilable viewpoints. They really don't know. Their opinions can't agree, since for want of agreement among them the object has failed to exist independently of them.

Schooled by my professors in the culture of objectivity (Norbert called my school the College of Unreason, in an allusion to Butler's book), I had trouble getting used to the idea that the Aramis actors were all telling us stories. At Orly, however, they had tried to tell a story *that held together.* To do so, they had gone ahead and developed bits and pieces in their workshop. They had gone to hangars, to catalogues, to manufacturers looking for electric motors, auto bodies, laser and ultrasound receptors, chips, concrete-and-steel sandwiches, feeder rails, electric circuits, emergency brakes, doors, and—to make it all work—activators with little electric motors. They really did try to bind all the different elements together. But Aramis didn't hold up. It turned back into a heap of disconnected scrap metal.

I had my own private interpretation, but I no longer dared express it to Norbert. Aramis didn't hold up because it was untenable. Period. From the very beginning, it was a fiction, a utopia, a two-headed calf.

Instead of that, my professor insisted—and with such arrogance!—that one had to live with variable degrees of objectivity! And wasn't it Unreason pure and simple that he was teaching me in *his* school? I had a name for his variable-geometry reality: I called it playing the sociological accordion!

No one can study a technological project without maintaining the *symmetry* of explanations.

If we say that a successful project existed from the beginning because it was well conceived and that a failed project went aground because it was badly conceived, we are saying nothing. We are only repeating the words "success" and "failure," while placing the cause of both at the *beginning* of the project, at its conception. We might just as well say that Nobel Prize winners are born geniuses. This tautology is feasible only at the end of the road, when we've settled down by the fire, after history has distinguished successes from failures. A comfortable position—but only in appearance, for as time passes the positions may be reversed: Aramis may become, in Chicago is going to become, the transportation system of the twenty-first century, and an obsolete VAL may be wiped out by the economic downturn in the Lille region. What does the pipe smoker say in such a case? Don't think for a minute that he's deflated: "Aramis was well thought out, you could see that right away. All you had to do was look at VAL to see that it was old hat." The incorrigible know-it-alls! They're always right; they always have reason on their side. But their reason is the most cowardly and servile of all: it's the one that flatters the victors of the day. De Gaulle after Pétain; Queuille after de Gaulle; de Gaulle after Coty. And it's the fine word "reason" that they distort for bootlicking purposes. *Vae victis,* yes, and tough luck for the vanquished!

No, honor and good luck to the vanquished. Failure and success have to be treated symmetrically. They may gain or lose in degrees of reality; one may become a utopian project and the other an object. This does not modify their conception, or their birth, or day 3, or day *n.* All projects are stillborn at the outset. Existence has to be added to them continuously, so they can take on body, can impose their growing coherence on those who argue about them or oppose them. No project is born profitable, effective, or brilliant, any more than the Amazon at its source has the massive dimensions it takes on at its mouth. Without modifying the explanatory principles, one has to follow projects lovingly through their entire duration, from the time they're just crazy little ideas in the heads of engineers to the time they become automatic trains that people take automatically, without thinking about them. Conversely, one has to stick with them while the automatic trains (on paper) that passengers (on paper) take as a matter of course turn back into wild ideas that float around, that have floated, in the heads of engineers. Yes, from the extreme of *objectivity* to the extreme of *subjectivity* and vice versa, we have to be capable of traveling without fear and without blame.

Let's not make a vertical separation between what exists and what does not exist. If we reestablish symmetry, then there is again transverse continuity between what exists and what does not exist, between VAL and Aramis. The project that does not exist is both easier and harder to explain than the one that does. To study VAL, classical relativism suffices—everyone has his or her own point of view on the thing; it's a question of perspective, of interpretation. To study Aramis, we also have to explain how certain points of view, certain perspectives, certain interpretations, have not had the means to impose themselves so as to become objects on which others have a simple point of view. So we have to pass from relativism to *relationism*. The war of interpretations is over for VAL; it no longer shapes the object; VAL's paternity, profitability, scope, maintenance, and appearance are no longer at issue. The war of interpretations continues for Aramis; there are only perspectives, but these are not brought to bear on anything stable, since no perspective has been able to stabilize the state of things to its own profit. Aramis is thus easier to follow, since the distinction between objectivity and subjectivity is not made—no "real" Aramis is the sum of the virtual Aramises—but it is also harder to follow, precisely since it is never possible to give things, as we say, *their due.* Everyone, even today, still tells us stories about it.

[DOCUMENT: TECHNICAL DESCRIPTION BY MATRA OF THE ARAMIS PROTOTYPE AT ORLY, JUNE 1973; EMPHASIS ADDED]

Report on the end of phase 0.

The Orly Aramis prototype: technical description.

The following pages are the outcome of a synthesis of the

reports and numerous explanations provided by Matra engineers during the Orly trials.

They should not be read as the description of a transportation system, but rather as that of a *prototype* of a transportation system—that is, of a *coherent* but essentially *changeable set* of solutions, often quite original ones, to the problems raised.

This "technical description" thus emphasizes the project's original features and neglects other more classical sets which have nevertheless been used and even sought after in the effort to avoid a "race toward gadgetry."

To study a technological project, one must constantly move from signs to things, and *vice versa*.

Aramis was an exciting discourse. It became a site at Orly. And now it has become discourse again: text, reports, explanations. What would Aramis' story be if it held together on its own? It would be a story that would work; it would carry away, would transport, without breakdown, those who gave themselves over to it. The door closes, the passenger punches in her destination, and the vehicle, without a hitch, without stopping at intervening stations, at 50 kilometers an hour, watched over from on high, delivers the traveler to her destination and opens the door. A whole program! But in 1973 Aramis is a *narrative program*, a story that is told to the decisionmakers, to stockholders, to local officials, to future passengers to "bring them on board," but it is also a work program, a flow chart, and a distribution of tasks, so that Matra can be an enterprise that works well. These programs are translated in turn by a computer program in 18-bit series: "111, give the number of the car; 0111, check the message; 01, open the door; 10, release the emergency brake; 00, display the number of the target station; 11, parity." Finally, all these mingled programs, all these trials, all these attempts produce a real story, written down: a ponderous text, *the report on the end of phase 0*. With Bardet, we were still at the point of daydreaming on paper, scribbling calculations on the back of an envelope; then we moved on to more serious writing, to patents. Next came protocols, signed agreements. Then we went on to hardware, to Orly, to grease and sparks and cement mixers and printed circuits. Now we've come back to print. The whole passage through hardware helps make the written history a little more credible. Bardet's affair was a tenuous dream: we weren't going along with it, it didn't work, they're pulling our leg. Lamoureux's story and Cohen's and Gayot's is a story that works: people believe in it, Orly is behind it—yes, unquestionably it all holds *together.*

The account of a fiction is generally easy to follow; you never depart from its textual form and subject matter. Wherever you look in the narrative of *The Three Musketeers,* Porthos, Athos, and Aramis always remain figments of the text itself. The account of a *fabrication* is somewhat more difficult, since any one of the figures may move from text to object or object to text while passing through every imaginable ontological stage. In order to follow a technological project, we have to follow simultaneously both the narrative program and the degree of "realization" of each of the actions. For example, the rendezvous of Aramis' platoons is an action programmed at the time of Bardet's earliest ideas, but its degree of realization varies according to whether we go from the earliest discussions with Petit to the patents, to Matra's plans, to the Orly site, to the imprinting of the chips, to the reports on the experiments, or to the report on the end of Phase 0. Depending on the point at which we look into the action, the "meeting of the branches," we will have ideas, drawings, lines in a program, trains running before our eyes, statistics, seductive stories, memories of trains running before the eyes of our interlocutors, photos, plans again, chips again. For the engineer *substitutes* for the signs he writes the things that he has *mobilized;* he *attaches* them to each other so they'll hold up; then he *withdraws* a little, *delegating* to another self, in the form of a chip, a sensor, or an automatic device, the task of watching over the connection. And this delegating allows him to withdraw even further—as if there were an object. If only we always went from signs to things! But we also go in the other direction; and we soon find ourselves not in a subway train but in a conference room, once again among signs speaking to humans—as if there were subjects!

Alas, VAL speaks well for itself, holds up all by itself. Why can't I? Oh, why did you never come to an understanding that would have endowed me with the same depth, the same weight, the same breadth as VAL? Why did you argue about me instead of agreeing on a unique object? Why was I words, and never the same ones, on your lips? Why can't I put identical words into your mouths? Proceed in a train! Move ahead! Split up! Behave! Go! Stop! Merge! Haven't I carried out all these orders? What more do you want? Do you want me to become embodied, to take on flesh? Action? That I am. Program? That I am. Verb I am as well. Why, oh why have you abandoned me, people? What do I have to do with prosopopoeia? Will you ever console me for remaining a phantom destined for a work of fiction when I wanted to be—when you wanted me to be—the sweet reality of twenty-first-century urban transportation? Why didn't you give me my part, the object's part? Why did Lamoureux, Gayot, and Cohen treat me as cruelly, as ungratefully, as Victor Frankenstein?

"It was on a dreary night of November that I beheld the accomplishment of my toils. With an anxiety that almost amounted to agony, I collected the instruments of life around me, that I might infuse a spark of being into the lifeless thing that lay at my feet. It was already one in the morning; the rain pattered dismally against the panes, and my candle was nearly burnt out, when, by the glimmer of the half-extinguished light, I saw the dull yellow eye of the creature open; it breathed hard, and a convulsive motion agitated its limbs.

"How can I describe my emotions at this catastrophe, or how delineate the wretch whom with such infinite pains and care I had endeavored to form? His limbs were in proportion, and I had selected his features as beautiful. Beautiful!—Great God! His yellow skin scarcely covered the work of muscles and arteries beneath; his hair was of a lustrous black, and flowing; his teeth of a pearly whiteness; but these luxuriances only formed a more horrid contrast with his watery eyes, that seemed almost of the same colour as the dun white sockets in which they were set, his shrivelled complexion and straight black lips . . . He held up the curtain of the bed; and his eyes, if eyes they may be called, were fixed on me. His jaws opened, and he muttered some inarticulate sounds, while a grin wrinkled his cheeks. He might have spoken, but I did not hear; one hand was stretched out, seemingly to detain me, but I escaped, and rushed down stairs. I took refuge in the courtyard belonging to the house which I inhabited; where I remained during the rest of the night, walking up and down in the greatest agitation, listening attentively, catching and fearing each sound as if it were to announce the approach of the demoniacal corpse to which I had so miserably given life." [From Mary Shelley's Frankenstein, or the Modern Prometheus]

"Do you comprehend the crime, the unpardonable crime?" Norbert asked me indignantly. "Victor abandons his own creature, horrified by what he has done. Popular opinion has got it right, because it has rightly given his name to the monster, who didn't have one in the novel. Frankenstein, all stitched together, full of hubris and remorse, hideous to behold. The monster is none other than Victor himself."

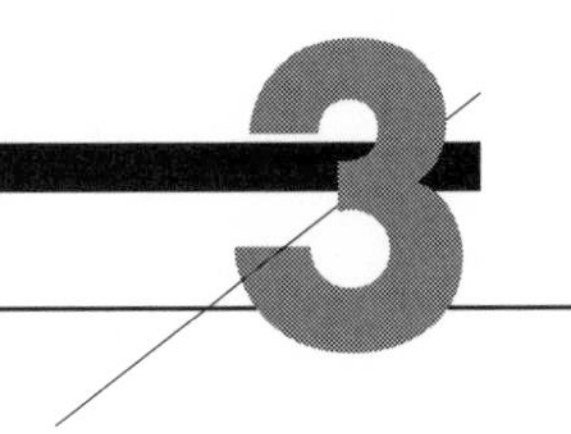

SHILLY-SHALLYING IN THE SEVENTIES

"We're making progress, my friend—we're crossing suspects off our list one after another. The cause of death can't be located in the final months; that much we've known all along. According to all the witnesses, it was inscribed in the nature of things. We know it can't be found in the initial idea, which everybody was excited about. We also know that it can't be found in the Orly phase, which went pretty well, all things considered, since it didn't commit the project to go in one particular direction or another. So let's move on to the other suspect phases."

"There's at least one point on which everybody agrees," I said, consulting the files. "After Orly, there was a period of shilly-shallying—"stop and go," they called it in Frenglish. You can see it in the chart of annual expenditures. Here: this goes from 1972 to 1987. I've marked where we are in black" [see Figure 1 in Chapter 1].

"You should have added the cost of our postmortem study in 1988."

"It's such a trivial amount, in comparison; it'd be invisible—a thin line at best."

"Yes, but if we'd done it five years earlier, they would have saved a small fortune!"

We have to overlook my boss's weakness for thinking he's useful and efficient, even though he himself of course criticizes other people's notions of usefulness and efficiency . . .

[INTERVIEW EXCERPTS]

At Matra headquarters M. Etienne is speaking:

"Aramis never had an engine. That was the real problem, *the congenital defect,* all those years."

"An engine?"

"I mean a local engine, a driving force: a politician, an elected official, somebody with pull who would have made it his cause, who would have put his shoulder to the grindstone, somebody devoted enough, stubborn enough . . . Somebody like Notebart, who supported VAL in Lille.

"Every public transportation system is so expensive that it needs a local political engine, a local base, to make it pay off over the years.

"But Aramis never had that. The godforsaken mess wandered from one end of France to the other and ended up in Paris, where, as it happens, there's nobody in charge of urban transportation. In Paris, it's the worst of solutions, and furthermore it had to fall under the thumb of the RATP, which is an enormous machine."

A technological project is neither realistic nor unrealistic; it takes on reality, or loses it, by degrees.

After the Orly phase, called Phase 0, Aramis is merely "realizable"; it is not yet "real." You can't use the word "real" for a nonfailsafe 1.5-kilometer test track that transports engineers from one beet field to another. For this "engineers' dream" to *continue* to be realized, other elements have to be *added.* So can we say that nothing is really real? No. But anything can become more real or less real, depending on the continuous chains of translation. It's essential to continue to generate interest, to seduce, to translate interests. You can't ever stop becoming more real. After the Orly phase, nothing is over, nothing is settled. It's still possible to get along without Aramis. The whole world is still getting along without Aramis.

The translation must be continued. What has to be done now is to recruit a "local engine" for this automated transportation system. And this engine in turn will attract local users, who will have to give up their cars and their buses in favor of Aramis. In order to oblige them, seduce them, compel them, Aramis will have to go exactly where they are headed; it will have to help them get there faster and more comfortably. Each one of us, by taking Aramis to reach our regular destinations, will contribute a bit of reality to this transportation system.

Obviously, Aramis won't find itself "up and running," ready to translate my crosstown itineraries, unless the local officials, after many others, become excited about its prototype and decide that Aramis translates their deepest desires. The task of making Aramis interesting never ends. For technology, there's no such thing as inertia. Here's proof: even an ordinary user can make Aramis less real by refusing to get into one of its cars; or, if she's a local official, by refusing to get excited about it; or, if he's a mechanic or a driver, by refusing to work for it. No matter how old and powerful, no matter how irreversible and indispensable, thus no matter how real a transportation system may be, it can always be made a little less real. Today, for example, the Paris metro is on strike for the third week in a row. Millions of Parisians are learning to get along without it, by taking their cars or walking. A few hundred shop technicians have stopped doing their regular maintenance work on the system, and a few dozen engineers who've benefited from the Aramis experiment are plotting to make the next metro completely automatic, entirely free of drivers and strikers, thanks to the Meteor project. You see? These enormous hundred-year-old technological monsters are no more real than the four-year-old Aramis is unreal: they all need allies, friends, long chains of translators. There's no *inertia,* no *irreversibility;* there's no *autonomy* to keep them alive. Behind these three words from the philosophy of technologies, words inspired by sheer cowardice, there is the ongoing work of coupling and uncoupling engines and cars, the work of local officials and engineers, strikers and customers.

So is there never any respite? Can't the work of creating interest ever be suspended? Can't things be allowed just to go along on their own? Isn't there a day of rest, after all, for innovators? No: for technologies, every day is a working day. You can forget the work of the *others,* but you can't manage if there's no one left working to maintain the technologies that are up and running. People who talk about autonomy, irreversibility, and inertia in technology are criminals—never mind the purity of their motives. May the ashes of Chernobyl, the dust of the Challenger, the rust of the Lorraine steel mills fall on their heads and those of their children!

[INTERVIEW EXCERPTS]

M. Parlat, project head, speaking at RATP headquarters:

"Before anything else, Aramis is a series of stop-and-go movements. That fact has played a major role. If you look at the history of the project, you see continual starting up and shutting down. Matra broke up its teams several times. How can you hope to get any continuity under these conditions?" [no. 2]

M. Etienne, director of Matra, responding to questions:

"Did this shilly-shallying make things hard for you?"

"No, we'd come to terms with the slow pace; the internal reason, too, which can now be revealed, was that we were supposed to keep the costs down. As early as '75 or '76, VAL had priority for us. Notebart had his shoulder to the wheel; we didn't have the same sort of pressure with Aramis—except from Fiterman, and that was later and didn't last very long.

"We didn't give up on Aramis, but Notebart kept after us. We handled the slowdowns without too much trouble, but it's true that in 1977 we almost entirely dismantled the Aramis team." [no. 21]

Messrs. Bréhier and Marey, speaking at the RATP bureau of economic studies:

"At Lille there was a real contracting authority. Here SACEM has one as well. But Aramis never had one. For SACEM, we have a need and we're figuring out how to meet it; for Aramis, we had an object and were trying to see where to put it, just as we've done with TRACS. But Aramis finally got finished."

"Finished off, you might say!" [laughter]

"There was a lot of fiddling around, a series of stops and starts that complicated things.

"At first, we thought we'd use the Bus Division, so it wouldn't get too heavy, like the metro (Aramis was supposed to be light, like automobiles). Then they said no, it was going to be in the Subway Division, since it was really tough work from the technological standpoint.

"The operating agency within the RATP was consulted only marginally. The operators didn't get involved in the project until the end, and in my opinion *it was too late.*

"The feedback loops between utilization constraints and technological constraints were relatively slow for the first ten years. At the same time, with a deal as innovative as this one, you can't let yourself be derailed by people who aren't used to it and who panic." [no. 30]

The time frame for innovations depends on the geometry of the actors, not on the calendar.

The history of Aramis spreads out over eighteen years. Is that a long time, or a short one? Is it too long, or not long enough? That depends. On what? On the work of alliance and translation. Eighteen years is awfully short for a radical innovation that has to modify the behavior of the RATP, Matra, chips, passengers, local officials, variable-reluctance motors—what an appropriate name! Eighteen years is awfully long if the project is dropped every three or four years, if Matra periodically loses interest, if the RATP only believes in it sporadically, if officials don't get excited about it, if microprocessors get involved only at arm's length, if the variable-reluctance motor is reluctant to push the cars. Time really drags. What happens is that actors get involved and back out, blend together or set themselves apart, take or lose interest.

Is VAL's time the same as Aramis'? No, even though 1975, 1976, 1977, 1979, and 1980 are critical years for both. It's no good taking out a chronometer or a diary so you can measure the passage of time and blame the first project for going too quickly and the second one for going too slowly. The time of the first depends on local sites, on Notebart's role as engine, on Ferbeck, and on Matra, just as the time of the second depends on the absence of sites, on hesitation over components, on the motor's fits and starts. All you have to do is reconstruct the chain of permissions and refusals, alliances and losses, to understand that a project may not budge for a hundred years or that it may transform itself completely in four minutes flat. The obsession with calendar time makes historians sprinkle technologies with agricultural metaphors referring to maturation, slowness, obsolescence or germination, or else mechanical metaphors having to do with acceleration and braking. In fact, time does not count. Time is what is counted. It is not an explanatory variable; it is a dependent variable that needs to be explained. It doesn't offer a framework for explanation, since it is an effect that has to be accounted for among many other, more interesting ones. Grab calendar time and you'll find yourself empty-handed. Grab the actors, and you'll get periodization and temporalization as a bonus. Here is sociology's sole advantage over history.

[INTERVIEW EXCERPTS]

M. Gueguen, the RATP's director of infrastructures, is speaking at company headquarters, in the cramped office he has occupied since the dismantling of the Aramis team.

"Aramis is point-to-point; it's on-demand service. That's the main difference between PRTs and all other transportation systems. There's no fixed line. But that part had to be given up right away, because of implementation issues. That was the whole problem for Aramis, in those ten years after Orly. The very thing that made Aramis so different, its ace-in-the-hole, was scuttled at the start.

"What costs the most in a guided-transportation system like Aramis that has an exclusive guideway? It's the infrastructure—the bridges, the tunnels, the viaducts, the tracks.

"Okay, Aramis has a big advantage in being small and light; but look, here's the problem. *[He takes out a piece of paper.]* They wanted to give good service, so they said: 'Let's put a station roughly every 300 meters.' Now in the metro, it's every 400 or 450 meters. Wait, though—the station is on a siding, because you don't want to block the cars behind that aren't stopping. That's the whole idea: not a single stop until the final destination.

"But the cars that stop have to have a shunt line to separate from the train. They also need room to slow down, plus a platform long enough to accommodate the number of cars, plus another shunt line so they can speed up and rejoin the train, which is running—or so Matra was saying at the time—at about 50 kilometers an hour without stopping. Do you see?

"Over a distance of 300 meters [Figure 6], if you don't want the passengers to be thrown all over each other (and passengers aren't cartridges or bottles) infinite acceleration and deceleration aren't possible . . . You see the problem?

"Well, you guessed it: there are actually *two* tracks almost the whole way. So even if Aramis is a narrow-gauge system, you have to carve out tunnels and make trenches for a very wide gauge! This is how an attractive advantage turns into a disadvantage. That's why the project changed shape so fast; it was impossible.

"So they said, 'Okay, PRT service is really two things: no intermediate stops, no transfers.' They kept the second, but they gave up the first. It's what's called a 'polystate.' Some stations are off-line, some aren't.

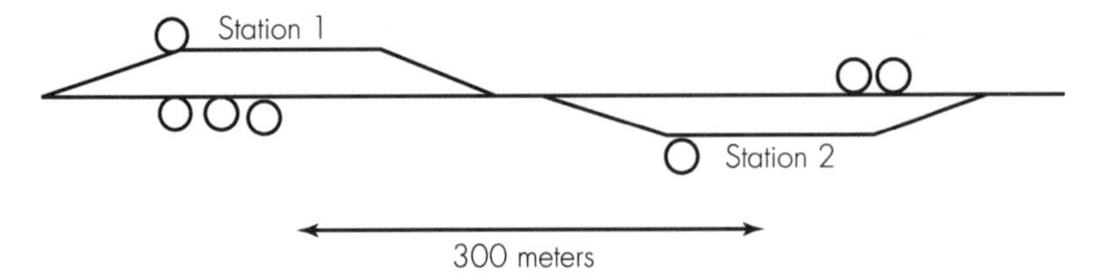

Figure 6.

"But at the same time, that had an impact on the idea of service on demand. At first, the idea was that it would be like a horizontal elevator. Every passenger presses one button to say he's there, waiting, and another one to say where he wants to go. The elevator decides on the best route. But a transportation system is always asymmetrical. People all go in one direction, then the other. If you let people direct their own cars to their destinations, at the end of the day all the cars would be at the end of the line; how would they get back?

"Or else you'd have to have so many cars, the system would have to be so enormous, that it would cost a fortune. So people said: 'We have to predirect the cars. The whole system has to be *controlled by the operating agency;* we won't need to do anything but post the destinations. The passenger gets into the car that displays the destination he wants.'

"But you can see that it gets complicated. If all possible point-to-point destinations are posted, there are problems with the itineraries, with the sign system. So we gave up user control of cars during rush hours; we kept it only for nighttime.

"So of the four ideas we started out with *[he counts on his fingers]*—one, passengers don't have to think, they sit down and only have to look up when they've reached their destination; two, they control their destination themselves; three, there are no intermediate stops; and four, there are no transfers, no changes—we kept the fourth one and waffled on the others: there are intermediate stops, but not so many."

"And moving as a train? What does that depend on?"

"Moving as a train is a matter of volume. If the cars are separate, with their security margins, there simply isn't enough volume. You have to have fixed groupings and give them considerable cruising speed."

"What's more, don't forget, is that the passengers are seated; that's an advantage, a strong point of the service, a plus that we kept almost to the end, but it's also, if you'll pardon the expression, a f—ing nightmare.

"For the RATP, remember, the only way to adapt transportation supply to demand is to take advantage of the compressibility of the human body. During rush hours, you compress people; that way the relation of supply to demand remains elastic. But this no longer applies to Aramis, since everybody is sitting down. If you're even one seat short, you have to have another car. *[Counting on his fingers again]* One, the problem of cost; two, the problem of system management.

"But even the polystate was so complicated to manage that Matra finally proposed a simplified Aramis, an omnibus that stopped everywhere.

"Of the idea we started with, only the elimination of transfers was left. The advantage is that we no longer had any stations on sidings, so we got back the single track with the advantages of a narrow-gauge system. We lost something, but we also gained something.

"But then that brought up a new problem. Since we now have omnibuses, there are fewer destinations, and in theory more people per destination; as a result, we can have larger cars. We started with four passengers, then we went to six, then ten, and then, after 1981, we had twenty passengers! Don't forget that the more people you have per car, the less it costs per seat. There was another advantage: it increased security, cut down on fear of crime.

"Only here's the thing: a car with more passengers but a narrow gauge is longer, and if it's longer it doesn't turn easily, and if it doesn't turn easily it has a hard time fitting into the sites. And don't forget that in the beginning, that was the promise Aramis held out: closely spaced stations, good adaptation to sites. It could turn in the narrowest streets.

"Well, the only way to make sharp turns is to have front wheels that can turn. It's like in supermarket carts. The front wheels go off in the right direction on their own, and that helps the onboard steering system. But if you have front wheels that turn, the car itself can't go anywhere but forward! And that's a tremendous constraint; they had the same problem in Lille with VAL. You have to have turnarounds everywhere to get the damn things pointed back in the right direction." [no. 10]

Projects drift; that's why they're called research projects.

To follow them, it's impossible to trace a target, a starting point, a trajectory. Point-to-point service by a programmable carriage taking two or four people without stopping along a network with multiple origins and destinations: that's the idea that got them all excited—Bardet, Petit, Lagardère, and their constituents. There it is: the goal for modern transportation. Now, this is precisely the idea that has to be given up first, because it is, their engineers insist, unimplementable. Let's say rather that it's "in contradiction" with other ideas that they also want to keep. Since these ideas conflict, they can't all be kept together within the same project. One of them has to give in. "Please, you go first." "No, by all means, after you." "Out you go!" Let's give up point-to-point and modify the goals.

Well, well! The engineers don't know what they want? The essence of technology isn't meeting objectives? There's a war of ideas going on in the heads of these peaceful engineers? There are internal contradictions? Trials and tribulations? That's right; because to relate means to ends is possible only in peacetime, when the engineers know where they're going and when they know in advance who agrees with whom, who allows what, who forbids or authorizes what. Now, Aramis is a research project because the engineers haven't yet completed their little sociology of ideas. Until the calculation was made, no one had noticed that stations on sidings every 300 meters would mean two continuous tracks and thus would wipe out the advantage of a narrow-gauge system. This is where the battle started. Locked up in the Matra engineers' craniums or computers, Station-on-a-Siding keeps Narrow-Gauge-System from enjoying its advantages, or else Speed has to shift from 50 kilometers per hour down to a snail's pace, or else the passengers have to turn into cartridges or pawns, or else the State has to give Matra an infinite sum of money, or . . . Everywhere, people are beating their head against the wall; no one wants to give in. Until the day when—but of course, the whole problem is those blasted stations! After all, why not give up Station-on-a-Siding? "Out of the question!" it yells indignantly. "Without me, Aramis is pointless!" Well, let's work out a compromise. We'll have one station out of three on a siding. The rest will be lined up omnibus fashion. That way we're holding on to the goal—though maybe it's a little tarnished now. The speed is still acceptable; the narrow gauge comes back into sight, and we can hope to transport flesh-and-blood passengers without knocking them about. Calm down, ideas! You see you've now become compatible. See if you can't get along together . . . until the next crisis. Which comes along very soon, when some sly fellow, taking advantage of Siding's demise, proposes to make the cars longer—they're now almost omnibuses anyway—so as to lower the costs of auto-

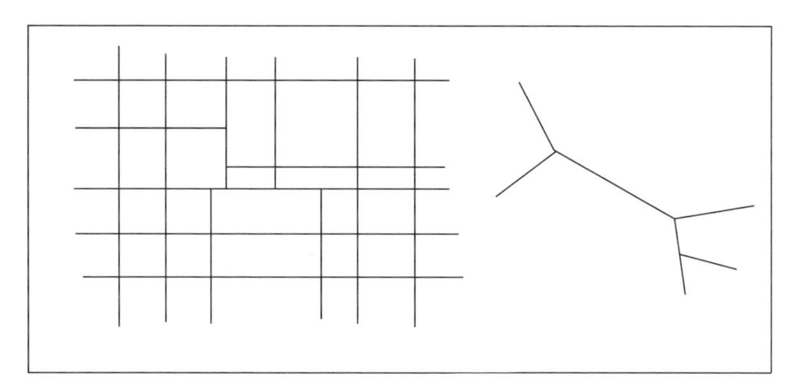

Figure 7. From a grid network to a branched system.

mation (they'll be divided by ten instead of four). "Out of the question!" exclaims Curve-Radius. "If I go from ten meters to fifteen, what will become of my buddy Site-Adaptation?" A new dilemma. New squabbles in the barnyard of ideas. They don't subside until Steerable-Front-End is introduced; however, this creature will agree to move in only if her fiancé, One-Directional-Car, comes along; and he'll cooperate only if his mistress, Complete-Loop-at-Every-Terminus, is part of the picture as well. That makes for quite a crowd; it costs a lot to keep them all fed, and they add up to a lot of problems for our engineers, whose initial aim was simply to produce Aramis.

"Watch out, now, this is a very important shift," Norbert said. "They started—on paper—with a dense network, lots of pickup points, small, reversible, four-passenger cars summoned by passenger demand and connecting all points of the network with no intermediate stops. They end up—still on paper—with a line on which nonreversible ten-passenger cars circulate omnibus-fashion. The passengers are still seated, it's true, and branching points allow the system to maintain a worthwhile volume at the ends of the line. Still, this seems to constitute an enormous transformation, a tremendous downgrading [Figure 7].

"But here's the hitch. In fact, in spite of the severe limitation of the network, all the problems of automation remain. The complete Aramis system is a little less problematic, but individual Aramis cars still have to be able to move together in a train, merge, demerge, and come together again. I have the impression that there's a fundamental problem here: no matter how much the implementation is simplified, the moving component remains as complicated as ever."

"I may be dimwitted," I remarked, "but even if we give up the agricultural metaphors I don't see how we're any better equipped to study innovations. I certainly learned how to calculate trajectories in school, but I had moving objects to work with, and degrees of speed, accelerations, measuring instruments, differential equations. How am I supposed to study Aramis, when there are no moving objects, no fixed points, when there's constant drift, when no one agrees with anyone else, and when I don't think you're in a position to come up with very many equations?"

"Stick to the actors, my friend, stick to the actors. If they drift, we'll drift along with them."

"But that's what you always say." It was actually one of my boss's little weaknesses: to stick to the slogans of his discipline no matter what, as if by repeating them he found himself less at a loss in confronting the technological arguments where we didn't understand a thing.

"Always simplify," he continued. "When things get too complicated, when the actors wear us down, we resimplify. After all, linguists describe all the sentences of a language through paradigm and syntagma; these are their words. 'The barber,' 'The barber left,' 'The barber left for the pond,' 'The barber left for the pond with a fishing pole.' You see, I'm following the syntagma, the syntagmatic dimension, the association, if you like; or, to make it even simpler, I'm using AND, as in programming language. But if you replace the barber with 'the grocer' or 'the butcher' or 'the programmer,' and if you change 'left' to 'returned' or 'arrived' or 'spit,' and if you replace 'for the pond' with 'for church' or 'for the oven' or 'for the mill,' and if instead of 'with a fishing pole' you say 'with a pen' or 'with a soldering iron' and so on, then you're descending, digging in; you're exploring all along the paradigmatic dimension. You're creating, setting up, the list, the series, the dictionary of all words and all expressions that could be substituted without altering the sentence's meaning, its plausibility. 'Spit,' for example, can't be substituted in this sentence, so it doesn't belong to the paradigmatic—that is, to substitution, or (if we simplify again) to OR, as in programming. AND and OR, association and substitution: it's as easy as pie.

"It's the same with Aramis. All the technological projects in the world couldn't make a set larger than all the possible sentences in a language. If linguists don't get discouraged, why should we? With these two dimensions, the paradigmatic and the syntagmatic, OR and AND, we can follow all the drifts. At first, everything goes along fine; sentence elements are added. 'Aramis is point-to-point,' 'Aramis is nonstop point-to-point,' and then, bang, you suddenly run into a statement that doesn't make sense: Aramis is feasible, but at infinite cost. End of the syntagmatic line. It doesn't hold together any longer. So you have to replace some of the sentence elements by other ones. You have to replace 'nonstop' with 'with some stops,' and 'stations every 300 meters' with 'polystate.' And that's how it works: you keep on drifting until you have a sentence, a project, that makes sense. You see, we're getting our bearings perfectly well, there's nothing to panic about."

Norbert took so much pleasure in enlightening me, he filled in his charts with so much satisfaction, that I would have felt guilty sounding skeptical. Student engineers (like engineers themselves, as I was soon to learn) can maintain a little peace of mind only if they respect the technological manias of their bosses and patrons.

In any event, I had learned something from M. Gueguen that was more important to me than that mishmash of paradigms and syntagmas. Every time I was squeezed in the metro at rush hour, I now knew that this was the RATP's way of adapting the supply of transportation to the demand. What gives an economic function its elasticity is the flexibility of my body! I found it easier to put up with the long rides our investigation required.

M. Etienne, head of Matra Transport:

"In 1974, the Transportation Ministry designated the RATP as the contracting authority for the project.* The RATP was flourishing at the time. Those were the days of Giraudet; the agency was modernizing the metro, doing away with ticket punchers (remember the ticket puncher at the Porte des Lilas station?), redecorating stations, installing an automated system to help steer the trains."†

"Giraudet was having so many problems at that time, with the reduction in the work force, the attitude toward change. Why did he put his weight behind a project as sophisticated as Aramis?"

"He had ulterior motives. Giraudet didn't believe in Aramis, but it didn't cost him very much to support it, and he had a good excuse: 'We're modern, future-oriented.' It didn't cost much; the RATP had the contracting authority for 250,000 francs. No, he got hold of the thing in order to scuttle it . . .

"When I came on the scene, it had already cost Matra 10 million francs. Once its feasibility had been demonstrated, we turned to the RATP and said, 'M. Giraudet, what are we going to do with it?' And he said, 'Oh, we're going to do the South Line.' That was its death knell.

"Giraudet knew perfectly well that it wasn't feasible. The thing meandered about, far from central Paris: there wasn't much traffic, there were fifty stations, maybe ten passengers an hour, and you had to plan on having 2,000 cars! After that, the RATP said, very astutely: 'Aramis is complicated. We're going to create something else—Arabus.' This was a site-specific bus system.

"Obviously, by doing comparative studies, they demonstrated that Aramis wasn't feasible. This is when we began to question the point-to-point feature, and the idea of preprogrammed control. This is what led to the idea of the polystate, at the time."

"Would Giraudet deliberately have led you down the garden path?"

"No, that's probably putting it too strongly, but he had enough insight to

*The "contracting authority" *[maître d'ouvrage]* is in charge of spearheading a project and has overall responsibility for it; the "contractor" *[maître d'oeuvre]* is responsible for carrying out the project, for its actual realization. Normally, the RATP also plays the role of contractor for its projects. It turns to industrial subcontractors *[ensembliers]* for specific tasks, but maintains full decisionmaking control.

†The Paris metro works on automatic pilot, in the sense that the driver's job is only to make sure everything is running properly. The only decision he makes is the decision to start up again, when he gives the order to close the doors. In VAL, as in Aramis and the Meteor, there is no driver at all, or, to be more accurate, the "driver" supervises the ebbs and flows from afar, from the control room.

ask for the South Line route, knowing full well it wasn't feasible. I was a little discouraged, and I didn't hide it. *[Pause.]*

"On the other hand, yes, he's quite bright enough to know exactly what he's doing. In 1980 I had a revelation: the southern part of the Petite Ceinture in Paris. I knew someone at the SNCF, Charles, who was responsible for the suburbs. He told me confidentially, making me promise not to use his name: 'I've got an idea. It's the Petite Ceinture, the PC. It's been closed since 1939; there's potential for traffic. It's not doing the SNCF any good, so maybe they'd let you use it.' I told the RATP (I'm pretty sure it was around 1980), 'Why not the Petite Ceinture?'" [no. 21; see the map in the frontmatter]

At the Institute for Transportation Research, M. Desclées and M. Liévin are sitting in a large sunny office overlooking lawns and fountains. In the distance, one can hear the rumbling of traffic on the turnpike leading south.

M. Desclées:

"The problem with Aramis is that they looked all over for a place to put it—Nice, Montpellier,* the southern suburbs—and they finally made the worst choice of all: inside Paris itself."

"Why was that the worst choice? I thought it was indispensable as a technological showcase."

"No, it's the worst, because there's a law I learned after years with the ministry: you can never introduce major innovations in urban transportation in Paris. For three reasons:

"First, because technologically speaking everything gets very complicated right away: there are always huge fluctuations in volume, and you're always right at the outer limits from the word go. You can't start small and expand progressively.

"The second reason is political. In Paris, you never know who's in charge: there's the mayor, the regional authorities, the prefect, the minister† . . . No-

*It was in Montpellier that the preliminary studies went farthest. Georges Frèche, mayor of Montpellier since 1977, a leftist and a great modernizer, wanted to implement a revolutionary public transportation system. The loop he wanted to insert in the old city was well adapted to Aramis, because it was not in competition with an existing metro. When the project was abandoned in 1987, Frèche wrote a virulent pamphlet in protest.

†Because of its size and importance, Paris is a very complex entity from an administrative standpoint. There are administrative agencies at the city level, the departmental level, and the regional level (the Seine Department is also part of the Ile-de-France Region). In addition, since Paris houses the national government, many administrative functions are carried out directly by a prefect at the State level. There are also numerous ad hoc organizations linking Paris with the suburbs.

body can commit himself; it's impossible. So by the same token there's never any support, never any local driving force.

"Third, in Paris you're necessarily dealing with the RATP, and the RATP isn't capable of undertaking any technological innovations where there's a high level of risk. This is not because their people are unenthusiastic, but because the RATP is a huge enterprise whose job it is to transport millions of people safely. So it goes overboard; they're perfectionists. They take so many precautions, they're so hypercareful, hyperprotectionist, that they can't take any risks, especially if the innovation is really on the cutting edge. And it's a good thing they don't take risks, actually; it's not their job."

"Sorry, but I don't understand. If what you're saying about the RATP is true, why did it latch on to the most complex, most innovative, most radical system, and why did it always oppose efforts to simplify the system, to make it sturdier, less innovative?"

M. Liévin:

"I think you have to start with the hypothesis that the RATP was *ambivalent* about it. I have a feeling that in the beginning the RATP wanted to sink VAL.

"VAL, VAL's success, Matra's success, was experienced as a tragedy by the RATP. It's the first time in France that a metro has been developed without the RATP, not to mention a fully automated one. High tech. The theory is a bit Machiavellian, but you can't rule it out."

"I'm probably being naïve, but I still don't understand. Why not produce a VAL, then, instead of Aramis?"

M. Liévin:

"No, on the contrary, the RATP couldn't get interested in VAL; VAL is much *too close* to the metro. What's more, VAL was earlier—it goes back to 1967. From the start, the RATP flirted with VAL in Lille. Some of its agents actually helped develop it in Lille.

"But VAL was a done deal; it had taken off. Whereas Aramis, while it was technologically more advanced, was *less destabilizing administratively.*

"They couldn't get interested in VAL without declaring that the era of the Paris-style metro, with drivers, was over. And they couldn't accept that."

M. Desclées:

"With a drivers' strike looming, and the whole nine yards . . . Yeah, Liévin is right: VAL confirms my three laws. It's in Lille. It's got the unanimous support of the politicians. It's not under RATP control. Besides, it's too close to being a metro. *Socially and institutionally,* that makes it unacceptable to the RATP. It's only now, ten years later, that they're talking openly in Paris about automating the metro. They've skirted the issue for ten years. At the time, it was unthinkable."

"Okay, I can see why VAL is unacceptable. But why choose Aramis in particular?"

M. Liévin:

"There was simply *nothing else on the market* at the time that was at all innovative or exciting.

"It *had* to be Matra, because the RATP *had* to get hold of Matra's technological competence, the competence they demonstrated in Lille.

"With Alsthom, they went way back, they didn't need an alliance. But with Matra, with a new gimmick, small-scale, local, which wasn't VAL . . . No, *it was the best possible alliance, the best compromise.*

"And beyond that, on the other side, the Matra-RATP marriage was very important to Matra; for them, it was a fantastic plug." [no. 11]

"We're onto something here," said Norbert. "For the first time things are getting a little clearer. The delegation of the contracting authority to the RATP in 1974 and the choice of a site inside Paris in 1980—these were both crucial decisions. They entailed commitments, they determined outcomes, they constituted the project's destiny."

I didn't dare point out to him that after each interview things "got a little clearer," only to get muddled again during the next one.

"Do you play Scrabble?" he asked me suddenly. "I'll play you for your salary."

The only way to increase a project's reality is to compromise, to accept sociotechnological compromises.

The good Scrabble player is not the one who uses permutations to get terrific words on his rack, but the one who succeeds in making good placements on the board, even if the words are shorter and less impressive. A few letters in a strategic position can bring more points than a fully spelled-out "Aramis" that you can't place anywhere and that forces you to give up your turn so you can keep it intact while you wait for a better configuration on the board. Besides, skipping your turn won't solve anything, since as the game progresses your chances of placing your word without any alterations or deletions may decrease instead of increase. Competitors proliferate. The board gets saturated. A wrenching moment for the engineer, as for the player entranced by his fine word: "Aramis" has to be *abandoned* for another combination, or worse still, has to be tossed back into the pool at the risk of drawing "zyhqhv"! It's this moral crisis that leads the pure Aramis—the first Aramis, the one that could do everything—to be called nominal,* while the series of altered and compromised Aramises is referred to as the *simplified* Aramis, or the *degraded* Aramis, or the *VS* (for *very simplified*) Aramis. If the player is reluctant to compromise his construction, he has lost. He'll be lucky if he can place the prefix "mis-," or the word "rat." Like the overly fastidious heron of the fable: "Hunger gripped him; he was quite happy, perfectly delighted / To come upon a slug."

The compromise is all the more difficult to bring about in that it really should blend social and technological elements, human and nonhuman agents. Behind the actors, others appear; behind one set of intentions there are others; between the (variable) goals and the (variable) desires, intermediate goals and implications proliferate, and they all demand to be taken into account. The engineers responsible for establishing Aramis in the southern suburbs have to take into consideration platform length, passenger flow, the number of mobile units, curve radii, and the fact that the cars are nonreversible. As if that weren't enough, the engineers also have to reckon with the age of the captain, as it were: the RATP director's ulterior motives, the corporate culture of the engineers, the history of disappointments built up by the VAL adventure taking place 200 kilometers away. The maintenance of Nominal Aramis requires so many sidings that the system ends up costing twice as much, it appears. But giving up Nominal Aramis would, apparently, infuriate the RATP engineers, who are looking for a new system that is as different as possible from VAL. Where is the technological component? Where is the social component? The only questions that count are the following: Who is compatible with what? Who agrees to stay with whom, and under what conditions, according to what hidden intentions?

*"Nominal" here means in conformity with the original functional specifications. The dictionary offers other, more conventional definitions that fit Aramis better—e.g., "existing in name only, not in reality."

"I embrace my rival—so I can crush him." In order to solve the problem of doubling the tracks without giving up stations on sidings, let's increase the distance between stations from 300 meters to one kilometer. Impossible, for Closely-Spaced-Service is furious and pulls out: How can you make passengers walk half a kilometer, when you're asking them to use Aramis the way they use their cars? Too bad, we'll just get along without Closely-Spaced-Service; let's increase the size. But if we abandon Small-Size, Aramis starts looking like nothing so much as an automated metro. All the RATP drivers, under the heading "Project Aramis," will read "wholesale automation of the Paris metro" as an entirely different project, "conquest of Paris by profit-hungry Matra, through an already-obsolete mode of transportation," and they'll all be furious to be treated as incompetents or underdeveloped types. After all, up to now haven't they been the best metro builders, the only ones?

But wait a minute, some will object, we're dealing with technologies, not passions; with drawings, not plots; with logic, not sociology; with economic calculus, not Machiavellian calculations. Ah, but they're wrong! The two sets come together in research rooms and administrative council rooms. The pertinent question is not whether it's a matter of technology or society, but only what is the best sociotechnological compromise. Neither the RATP nor Matra can agree at the time on a mini-VAL inside Paris,* but they can agree on an Aramis that must be neither too simple (or it will infuriate the RATP) nor too complex (or it will cost a fortune). It must be neither too far from Paris (or it will infuriate Matra and run in the red) nor too close (or it will have too many passengers and be burdened with too many tedious regulations). This is illogical? No, socio-logical. Aramis inside Paris is an alliance, a compromise. The "best possible alliance," said our interlocutor, M. Liévin, a modern-day Pangloss. Let's say instead that there are as many possible Aramises as there are possible compromises among all those—humans and nonhumans—who have made themselves necessary to its gradual realization. The only impossible solution is an Aramis that would accept no compromises; that would suspend the work of recruiting, of generating interest, of translating; that would expect the Orly Aramis to come into being all by itself, on its own power, from the feasible prototype

*Even though the Meteor project today constitutes a new compromise between Matra, VAL, and the classic Paris metro, since it is automated like VAL but with the volume of a metro. As we shall see, some observers credit Aramis for this compromise, which it finally made acceptable.

to the real transportation system, as if the words in a Scrabble game jumped all by themselves from the rack to the board, as if the characters in a novel could go out in the street without their readers, as if the monsters of Victor Frankenstein's laboratory could simply skip town.

"Well, Norbert, are your syntagmas and paradigms still working?"

"Of course, chum, of course, how can you think they might not be? Studying a technological project isn't any harder than doing literary criticism. Aramis is one long sentence in which the words gradually change in response to internal contradictions imposed by the meaning. It's only a text, a fabric. Look:

Established in Orly.	2 passengers in 1971.
Established south of Paris.	4 passengers in 1972.
Established in Marne-la-Vallée.	4 passengers.
Established at La Défense.	4 passengers
Established in Montrouge.	6 passengers in 1974.
Established in Toulon.	6 passengers.
Established in Dijon.	6 passengers
Established in Nice.	10 passengers in 1977.
Established in Montpellier.	10 passengers.
Established on the Petite Ceinture in 1980.	10 passengers in 1982.
Established on Victor Boulevard in 1984.	20 passengers.
Established nowhere in 1988.	0 passengers in 1988.

The passengers are seated in 1973.
The passengers are seated.
The passengers are seated.
The passengers are seated.
Some passengers are also standing in 1987.
There are no passengers in 1988.

"There's the declension of our Aramis, the paradigms. And here are the long sentences of the Aramis story, which hangs together better and better. Here are the sequential syntagmas:

"1. Aramis is a programmable bench that interests all city-dwellers and especially M. Petit of DATAR;

"2. Aramis is a machine transfer system that obeys the laws of kinematics and interests M. Bardet;

"3. Aramis is an original means of transportation that follows the fashion of PRTs and that interests Matra, which is attempting to diversify and which is applying automobile and space technologies to guided-transportation systems;

"4. Aramis is a small car that has four yellow seats and electronic couplings, that can link up, separate, and merge, that has probabilistic security, that is tested at Orly, and that is financed by Matra, the Transportation Ministry, Aéroport de Paris, DATAR, and the RATP;

"5. Aramis is a revolutionary transportation system that has 2,200 nonreversible cars with six seats each, onboard steering, and sixty stations on sidings, that allows passengers to go where they want to go without transfers and without intermediate stops, that serves the southern suburbs, and that costs a fortune."

(Gasping for breath like a singer.)

"6. Aramis is a new transportation system that has cars with ten seats, that is equipped with variable-reluctance motors, that uses the omnibus format, with stops but no transfers, that has branching points requiring separations, mergers, and couplings, that serves the Petite Ceinture in the south of Paris, that has the RATP as its contracting authority.

"And so on.

"The story gets longer and longer. It's a sea serpent, a more and more complex sentence, but one that's more and more reasonable, since through trial and error it has rejected or eliminated everything that didn't make sense. It becomes so complete, so comprehensive, so enveloping, so detailed, that volumes of reports and specifications are needed to contain it."

Since the 1960s, French intellectuals have had a certain weakness for seeing texts everywhere. For my mentor Norbert, educated in the era of Barthes and Lacan, Aramis was becoming literature; it was a poem to be composed by modifying sentence components, trusting to luck, as with Raymond Queneau's 100,000 poems.

As a young engineer, I refused to confuse motors and railroad tracks with words.

"But which one of those sentences tells the right story?"

"All of them, none of them. The one you get carried away with; the one that would really carry you away, from the boulevard Victor to Bercy—that's the one that would be the right story."

"But then the right story would be endless; the sentence would be so long you wouldn't have enough breath to say it, or enough paper to write it down."

"No, because the more reality we take on, the more the arithmetic changes. An infinity of stable elements becomes a single element. For you, for the user, the sentence becomes: 'I'm taking Aramis and I'm on my way.' Not even. The sentence is: 'I'm on my way, I'll be at the boulevard Victor at 4:30.' You don't even mention Aramis' name any more."

A text so long that it becomes a mute thing, a text so long that it ends up recruiting enough things and endowing them with enough meaning to keep quiet and make me speak, me, the animal endowed with speech—this mystified me even more than the puzzle we were supposed to be solving. Now that my mentor no longer dazzled me, I'm sure the reader will understand why I took steps at that point—unsuccessfully, alas—to change my internship and enroll in the program in Man-Machine Interaction and Artificial Intelligence.

[INTERVIEW EXCERPTS]

M. Berger, speaking at RATP headquarters:

"Throughout this whole period, you must remember, there's something that works really well, it's a major innovation: I'm talking about the rotary motor, the variable-reluctance motor. It isn't functional; it's not related to the transportation system; it's not electronic. But maybe this is precisely why it's a great success." [no. 14]

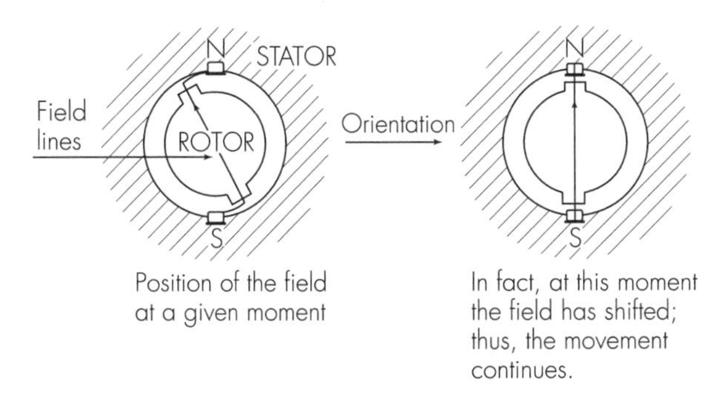

Figure 8.

"We still really need to understand what this rotary motor is all about. Everybody says it was Aramis' biggest success. You're a real engineer, so that's your area, chum; it's your job to understand that."

"According to what I've been told, it's an electric motor that doesn't involve a power transfer. You don't have a shaft or a sprocket; there are no mechanical parts except for the rotor, which is the axle. You have little notched wheels that allow very finely tuned displacements" [see Figure 8].

"Okay, we can leave it at that. Anyhow, it's a black box that looks pretty isolated in the project. If nobody goes into it, we don't need to go into it either. It's independent. It's not what scuttled the project."

You can see that my boss transformed his rules of sociological method into convenient devices for absolving his own ignorance and his intellectual laziness . . .

[DOCUMENT: CONCLUSION OF THE CET REPORT]

The motorization of Aramis with the variable-reluctance motor, which has been homologated by the RATP, represents an important innovation as a new mode of propulsion.

[INTERVIEW EXCERPTS]

In M. Hector's office at the Research Bureau of the Transportation Ministry.

"Still, it wasn't a complete waste; there was the variable-reluctance motor."

"Yes, but do you know a single application for it? Do you know one person who wants it?"

"No, you're right. In robotics, though, don't they think it might be useful?"

"Does it have the right power?"

"Uhh, no."

"So you see, even the motor, okay, it works, but nobody wants it. Can you call it a success, if there are no takers?"

"Okay, I see your point. Anyway, you can't use the motor to justify the whole Aramis project. Five hundred million francs for a motor nobody wants—that'd be pretty expensive."

"Right!" [no. 31]

In the first place, a project isn't *one* project. It's taken as a whole or as a set of disconnected parts, depending on whether circumstances are favorable or unfavorable.

In the first place it's abstract,* since each element, once drawn in, once "interested," pursues its own goals and tries to conform as little as possible to the common translation. It becomes concrete only gradually, if it can count as one in the eyes of all its users, who are simply content to "take Aramis" on the Petite Ceinture so they can go about their business. But if circumstances are against it, then the project decomposes, and each element, disinterested, goes its own way, making Aramis more and more abstract. That's why the sum of the elements that constitute it can never be fixed once and for all, for it varies according to the state of the alliances. In 1971 Aramis is a programmable bench allowing point-to-point service with on-demand user control. The word "Aramis," the character called Aramis, seems to be unique and unified. It really does designate a project, an idea, a concept. Yes, but it is still no more than an idea, not very real, not very realistic, not very fully realized, and it interests very few people.

*In Gilbert Simondon's sense; see *Du mode d'existence des objets techniques* (Paris: Aubier, 1958; rpt. 1989).

In 1976, on the other hand, it is composed of hundreds of *membra disjecta,* each of which is evolving in terms of its own alliances and prohibitions. The variable-reluctance motor holds up, but not the stations on sidings; and not the polystate. Matra sticks with it, but not Aéroport de Paris, which has long since dropped the whole thing. Finally, in 1988, there's no longer anyone who can explain what constituted Aramis' unity, since it has been dismembered once again, since the teams have been disbanded and the research center destroyed, and since there's talk of "spillover": the variable-reluctance motor is now isolated, abstracted from the whole, saved from the shipwreck to become one of the positive elements of Aramis endowed retrospectively with a life of its own. "At least there was the motorization—a great technological success, even if it's a little costly and hasn't yet found clients."

To count these successive states, a very peculiar arithmetic is required, since each element, by virtue of the pressure brought to bear on it, can become either an autonomous element, or everything, or nothing, either the component or the recognizable part of a whole. The question of how many elements compose a technological system cannot be answered by ordinary arithmetic. When there are breakdowns, accidents, strikes, the components separate into individuals and proliferate; then, when the system starts up again, they disappear, and literally do not count any longer. The lack of flexibility of the space shuttle Challenger's O-rings after a cold wave became apparent only after the explosion—from that point on, it was necessary to reckon with cold temperatures.* The lack of flexibility of the maintenance-shop workers in the Paris metro after an action by the CGT [Confédération Générale du Travail, a labor union] became apparent only after the long strike in 1988—from that point on, the workers had to be reckoned with. Before the catastrophe, didn't it ever get cold at Cape Kennedy? Before the strike, weren't there any bitter young nonunionized workers? The cold and the young workers were absent and present at the same time. Yes, it's a weird arithmetic that can't pin down the numbers.

*Thomas F. Gieryn and Anne E. Figert, "Ingredients for a Theory of Science in Society: O-Rings, Ice Water, C-Clamps, Richard Feynmann and the Press," in S. Cozzens and T. Gieryn, eds., *Theories of Science in Society* (Bloomington: Indiana University Press, 1990). See also J. L. Adams, *Flying Buttresses, Entropy and O-Rings* (Cambridge, Mass.: Harvard University Press, 1991).

[INTERVIEW EXCERPTS]

M. Parlat is speaking in the small, unused shanty on the boulevard Victor:

"We hesitated a long time over the transmissions. We started with optical linkages for short distances and ultrasonic ones for the long distances. The couplings between cars went through a lot of changes. In 1974 we had ultrasound for long distances and optical couplings for short distances; then, in 1983, everything was ultrasonic; finally, we had hyperfrequencies for long distances and ultrasound for short ones." [no. 2]

The topology of technological projects is as peculiar as their arithmetic.

If a project were made up of Russian dolls, everything would be simple. You would take the big doll, knowing that there were other ones inside, but you'd let the experts—who are more and more highly specialized—open up the dolls, one at a time. So you'd have a transitive series of dolls, each of which would hide and protect the next one down, and also a transitive series of specialists, each of whom would hide and protect the next one. You want to know how the rotary motor works? Eighth door on the left after you get out of the elevator: see the assistant head of the suboffice of motorization. You want to talk about a site for Aramis? That's the big doll that contains all the others, smiling, friendly, attractive. Go see the director, the fortunate M. Etienne. Division of labor, division of problems. A good flow chart to handle all the embeddings, and we'll be all set. A technological project would thus be neither complex nor simple; it would be a well-organized series of Russian dolls.

Unfortunately, arrangements of this sort are valid only for finished objects that need simply to be supported and maintained. The lovely series of successive tasks and embedded specializations is not valid for projects. The dolls aren't Russians but savages, or Russians like the salad of the same name, or like the nationalities liberated by perestroika. Their topology is so nonstandard that the very smallest one may, from time to time and for a certain period of time, contain the larger ones. Is nonmaterial coupling Aramis' content or its container? Hard to say. If Aramis were in operation today, its transitivity would be easy to observe: Aramis would be the transportation system of the Petite Ceinture, which would control a dense network, which would control detachable mobile units, which would control nonmaterial couplings, which in turn would control communications and programs for linking up. From the general to the particular, the logic would be easy to follow. But in the real Aramis, precisely the one that does not exist, we see that nonmaterial coupling becomes, for some people, the big doll that contains and justifies all the others: "It's the biggest innovation in guided-transportation systems." The dense network is only one of its consequences; as for the Petite Ceinture, it is only an unimportant experimental site. From footsoldier to general: the chain of command is clear, but reversed.

For projects or objects that have broken down, this sudden mutation may take place at any point in the old chain. Is the eighth microprocessor that controls the movement of trains in its turn contained by or contained in the nonmaterial coupling? Here's a new uncertainty. And it grows larger as it's examined carefully, to the point of blocking the entire Aramis project. From two cars to three, from three to four, the logic is no longer so sound. There's no possible generalization. At the third car, the train is disturbed, starts bunching up like an accordion, gets dislocated . . . The last doll becomes the first one, the one that holds all the others: the nonmaterial train doesn't hold up more than a few seconds.

"Engineers think like savages, my friend, as Lévi-Strauss does not say. It's a matter of tinkering with what you have on hand to get yourself out of terrible muddles: what was only a stage becomes an infinite number of stages, a real labyrinth. What was once a command chain becomes an unruly mob, a true case of every man for himself."

"But wait a minute," I exclaimed, indignant at so much bad faith and because, by chance, I had read Lévi-Strauss for my exams. "Lévi-Strauss contrasts modern engineers with mythical tinkerers.* We engineers don't tinker, he says. We rethink all programs in terms of projects. We don't think like savages."

"Hah!" Norbert muttered ironically. "That's because Lévi-Strauss

*Claude Lévi-Strauss, *Savage Mind* (Chicago: University of Chicago Press, 1968; orig. pub. 1962).

did his field work in the Amazon rain forests, not in the jungle of the Paris metro. What he says about tinkerers fits engineers to a T, his ethnologist's bias notwithstanding."

"But there really are experts in the field," I replied, letting my exasperation show, "engineers who don't tinker, who don't compromise, who don't translate, whose instrumental universe is not closed, who . . ."

"Of course there are, my friend," Norbert replied calmly, "when everything is running like clockwork, when the project is over, when everything is going along swimmingly; of course, then it's as if there were 'experts' quite unlike tinkerers and negotiators. But at the end, only at the end. And since Aramis wasn't lucky enough to have such an end . . . No, believe me, you don't have those who tinker on one side, and those who calculate on the other."

[INTERVIEW EXCERPTS]

M. Girard, who has since left the RATP, but who was responsible for Aramis' fate when he was chief executive officer:

"Aramis always suffered from *being badly integrated* into the RATP culture. When I was in the Rail Division of the RATP, I wasn't the least bit motivated by Aramis, whereas I did all I could for AIMT—the wholesale automation of subway trains . . .

"Aramis was interesting because of its rendezvous and its dense network, with multiple origins. I had suggested to the Bus Division that they should take over the project. It's lightweight; it's automobile technology, in terms of its construction; it's on tires. We had pulled out a few engineers, but there again *the RATP culture was hostile.*

By simplifying it, we denatured it; when we abandoned the point-to-point principle, we *killed the project* right there, and those who simplified it *didn't realize* the consequences their *decisions* would have." [no. 18, p. 4]

"Here's the first trail, my dear Watson, the first serious trail. 'We killed the project by denaturing it, by simplifying it.' They shouldn't have abandoned point-to-point, the dense network."

"But everybody else says just the opposite: 'The thing was doomed because it wasn't simplified enough, it wasn't fleshed out enough, it wasn't changed enough, there was no local force driving it."

"That's it: we're getting hot, we're closing in on the major argument about what is negotiable in a project and what isn't. One group says that the initial idea had to be abandoned right away because it wasn't feasible, and the other guy claims that giving up the initial idea is what did the project in."

"Guess what I just found in the *Larousse Encyclopedia!* An article on Aramis! A long paragraph—listen to how it ends: *'Adoption of the system was considered on the basis of studies undertaken by several French cities, Paris in particular.'*"

"Well, well! If your name appears in the *Larousse Encyclopedia,* you exist a little bit after all."

[INTERVIEW EXCERPTS]

M. Etienne, head of Matra:

"Some good engineers convinced us—and they were right—that you had to proceed in stages, starting with subsystems.

"Michel Frybourg, head of INRETS at the time, had convinced everybody, myself included. It was a reaction against the Bertin style of invention: Bertin invented a whole system around a single component, the linear motor, and it was a dud.

"So we said, 'Okay, let's do it in stages.' We knew, for example, that at Orly we didn't have the right motor; that's when the idea of the variable-reluctance motor came up, with the Jamet brothers.

"We did the bulk of Phase 1 on the motor. We said to ourselves, 'We're not going to try to implement the system until we have the components and the subsystems.' *Everybody was encouraging us to work in terms of subsystems.* They were telling us, 'We won't come on board if you move too fast.' Everybody agreed, and it wasn't stupid; I was as convinced as anyone, and I still am.

"Maybe we manufacturers are in more of a hurry than everyone else. But we were moving slowly. Everybody was encouraging us to proceed slowly. 'Watch out! Watch out!'"

What counts, in a technological project, is deciding what has to be negotiated, and deciding on an official doctrine that will make it possible to proceed with any negotiation at all.

To hesitate is not necessarily the same as to renegotiate. In Scrabble, too, a player may hesitate before placing the perfect word that she's composed so lovingly on her rack. She can even break it in two and resign herself to placing only half of the word—for example, "Ara" ("climbing bird—Platycercides—scientific name 'sittace,' large South American parrot with brilliant plumage") instead of "Aramis" or "Ararat" ("volcanic peak in Turkey where Noah's Ark landed after the Flood"). But hesitation and breakup do not lead to the same solutions as does the renegotiation of the whole word according to the positions to be occupied on the board and the letters available on the rack. Intellectually and emotionally, the two operations remain different, even if, for the other players, they are both translated by a period of anxious waiting and a lot of head scratching. Should we hesitate to implement Aramis, or should we completely rebuild it?

At this point, between 1974 and 1980, nothing yet proves that "Aramis" is a good word. Nothing yet proves that it should not be completely reformulated. However, the people carrying the ball for the project make a major decision as to what is negotiable and what isn't. Implementation, size, operation, financing, the dense network—all these are open to discussion; the mobile unit with nonmaterial couplings is not. Now, such a decision constitutes a commitment for the future, since it brings to the negotiating table players who don't have freedom to maneuver—freedom that they'd need in order to be able to rethink the entire project as circumstances change. They're obliged to maintain the mobile unit intact (it has become the heart of Aramis, its essence), whereas everything else is subject to change. Now, how can one make a clean break, in a transportation system, between the unit that moves and the network in which it moves and the city in which the network is to be implanted? Yet the decision isn't really a decision, since it remains implicit. No one entrusts the negotiators with a list of priorities and a margin of maneuverability for each item; furthermore, no one recognizes that there is negotiation going on, with a negotiating table, priorities, and *maneuvers* to be carried out successfully. On the contrary: in place of such a list, we find a doctrine of "common sense," a piece of strategic advice offered by M. Frybourg, who recommends carefully distinguishing between finding sites for the system and perfecting its components, such as the motor, the mobile unit, the transmission systems. This hard-won "wisdom" acquired after the failure of a few grandiose projects—like Bertin's aerotrain—may turn into "folly" in Aramis' case. No retroaction is going to make it possible to redefine the mobile unit in light of variations in implementation or operation.

People who study technological projects take too little interest in the official doctrines dealing with the actual management of the projects. This metalanguage appears parasitical. Yet it plays the same essential role that strategic doctrines play in the conduct of wars. In the course of a battle or a project, *ideas* about the way to handle battles or innovations play a performative role. To separate the perfecting of the components from the perfecting of operating conditions and sites, in the name of the hard-won wisdom acquired in the board room of the Department of Bridges, is to make a decision as important for Aramis as the decision to reinforce the Maginot Line in the name of the hard-won wisdom acquired at military headquarters. Writing a project's history also means writing the history of the ambient theories about project management. The history of technologies has to include the history of these doctrines, just as military history includes the history of the strategies taught in war colleges and at general headquarters.*

"But engineers don't play war games, Norbert. In fact, they don't play at all; they work. They're not 'sent to the negotiating table'; they perfect the most effective technology possible. In fact, they don't 'negotiate' at all; they calculate. They don't pay any attention to 'changing circumstances'; they leave that up to the politicians and the marketing strategists and other busybodies. What engineers do is think about the optimal solution. They have no use for 'priority lists' or 'maneuvering room,' or all the rest of the hoopla thrown around by rhetoricians and business schools. Don't forget, engineers aren't trained at Sciences Po [Ecole des Sciences Politiques]. They're honest and upstanding, and they don't give a hoot for the 'doctrines' of merchandising and compromise."

*For military examples, see J. Keegan, *The Mask of Command* (New York: Viking, 1987).

"That's an interesting theory all right, and it's often come in handy," Norbert replied ironically. "In fact, that may be exactly what killed Aramis. It would give us a pretty good criminal."

[INTERVIEW EXCERPTS]

At the Institute for Transportation Research, M. Liévin, affable and sarcastic, is speaking once again:

"There's the main difference, I might say, between Aramis and VAL. VAL, in Lille, started small, and gradually got complicated in accordance with local demands.

"At first it was referred to as 'the whatchamacallit.' That should give you an idea. I was there when it started up, in 1968–69. Watch it, now, don't go and call me one of VAL's fathers; it already has a lot of fathers."

"Yes, it's not like Aramis. There aren't any paternity battles over failures!"

"Exactly. Okay, so we did the studies and we were dealing with an *internal* transportation system in Villeneuve-d'Asq.* Ralite, an engineer from the General Council on Bridges and Roads, asked us to do a study; he really took us by surprise the day he asked us to extend the system so it would go all the way to the Lille station.

"We were very much influenced by a Swiss mini-train, the Habegger, which still exists. Then Ralite gradually presented his project as the beginning of the Lille metro.

"A call for bids was put out; Notebart had assured us that the urban community would go along.

"In 1972, when the bids were examined, there were five proposals: Matra, Bertin, Soulé, I can't remember the others. It should all have been automatic.

"There was a meeting. The Matra bid was officially declared the best.

"At that point Professor Gabillard, from Lille, a hyperfrequency specialist, and very active on the local scene, a well-known figure, a Rotarian, sensed that the university could profit from the situation, and he latched onto the idea. He hasn't let go, either; he's still there. You really have to go see him.

"He put the Habegger and Matra's proposal together. His ideas have been twisted around several times, but he was a very effective stimulus. The papers

*Villeneuve-d'Asq was a newly created town on the outskirts of Lille where a new university was being established. A novel transportation system was to be added to the novelty of the town itself.

were full of articles on the 'Lille inventions.' In terms of local implementation, we couldn't have done better; and besides, we moved from something simple to something complicated." [no. 15]

"If I understand correctly, to write the Aramis story we have to rewrite the VAL story."

"Yes, since it's the same company, Matra was working on both systems during the same period, and the two project heads had been classmates. Furthermore, this gives us an admirable point of comparison: a failure, a success, the same men. There's no way to say the industrialist is no good, since he succeeds in one case and loses in the other. With the comparison, we're compelled to see the symmetry. And technologically speaking, one is as complicated to produce as the other, even if Aramis is even more on the cutting edge. No question about it: I sense that we're about to add a chapter to Plutarch's *Lives,* and if you still think it's impossible to be both a good engineer and a good negotiator, get ready to squirm, Mister Young Engineer."

"But I've been squirming all along, Mister Literary Graybeard!" As you can see, over time I had become rather cheeky.

"Right. So our next meetings will be about VAL?"

[INTERVIEW EXCERPTS]

At Matra headquarters, in the office of M. Frèque, who worked on VAL before going back to Aramis in 1981. The same postmodern ceramic building. The same charming hostesses. The Aramis cabin is still in the hall. Delighted to be able to talk about his offspring VAL again, M. Frèque compares the two projects for an hour, without waiting for questions.

"Yes, it's true, the contrast between Aramis and VAL is striking. In our response to the call for bids, we had proposed a single line without a network. We had nonreversible units with loops and a door at each side, for the purpose of simplification. The functional specifications of the competition were *very few in number:* a one-minute interval during rush hours, and a fairly low external noise level, for the site. The greatest possible technological *freedom* was left to the manufacturer. That was in total contrast with Aramis.

"With VAL, we moved gradually from something *simple* to something *complex.* The system underwent two major changes in definition, one functional and one technological.

"The most important by far occurred in 1974, after the trial runs on the test track with the prototype—there was a CET, just as there was for Aramis.

"Notebart showed up—he hadn't shown much interest in the project up to that point—and said, 'It's a mission for the future, it's a very good image on the political front, it's the rebirth of public transportation. A route in the new city is great, but if we used it to connect the new city and Lille itself, it would be better yet. What I want is a *network.* What do you have to *change* in your whatchamacallit to do a network?'*

"For us it was an extraordinary opportunity. He had taken on an adviser, Rullman, an Ecole Polytechnique graduate, who played a very useful role; he was an RATP retiree. At first we took him for an old fart, it's true. We were young; we said, 'Why's he hanging around bothering us with all his old-fashioned ideas?'

"Still, the *confrontation of ideas* got fully played out. It was a crash course! Actually, he was quite independent-minded and he often disagreed with the RATP; those guys didn't believe in wholesale automation. He was a good guy, Rullman, because he always said, 'You have to . . .' That's what the experts always said: 'You have to . . .' I always looked for technological justifications behind their 'You have to,' to see if it might be just a question of habit.

"So we had serious hassles because the shift from the little line to the network didn't go smoothly. We had to go from *nonreversible* to *reversible* so we could extend the lines, and that led to really major changes.

"The contracting authority for the new city stuck with us; that was Ficheur, a terrific guy, really dynamic, who unfortunately died quite young.

"At first the people of Villeneuve didn't like the idea that we were *complicating their system,* because that increased the costs and the time frame, and then the specifications changed. But Gabillard, who was against it at first—he always said, 'You're going to complicate things'—finally came out in favor of the change, and from then on all the others followed along.

"Within six months, before the end of 1974, everything on the test line had been redone. I remember, it was before the end of December . . . There were violent arguments, but in six months' time we had adjusted to the specs. The

*On the administrative history of the project, see the somewhat self-serving book by Arthur Notebart, *Le Livre blanc du métro* (Lille: Communauté Urbaine de Lille, 1983), 190 pages with photographs.

system was a little more complicated: it moved from extreme *simplicity toward increased complexity.* That, too, is the opposite of Aramis.

"Be careful here, you mustn't think that the project developed differently from Aramis because it was less innovative. VAL was just as new, at the time. In Atlanta,* Westinghouse had produced an automated metro, but the distances were very short, and they'd also put in a moving sidewalk in case of breakdowns! And it was all in tunnels, whereas we had both viaducts and tunnels. That may not seem like a big difference, but it actually changes a lot of things.

"No one believed in VAL. It was a first. The RATP people had said, 'We don't want our name involved in a thing like that.' They even wrote a letter—obviously they regret it today—in which they said they could take no responsibility whatsoever if wholesale automation was involved.

"People were saying, 'It's a new Concorde,' 'a new aerotrain.'

"Everyone agreed that it wouldn't work. Notebart's opponents were the ones saying that. 'VAL is kaput,' as they claim up there. I have to say that in transportation, there have been quite a few innovations that have failed.

"The second major change came about in 1977. I noticed that the CMD for a minor increase in performance brought the cost up tremendously. I proposed to the Lille authorities that we should modify it after the contract was signed. I'd thought a little about it before, but I didn't want to stir up any trouble!

"It was a minor downgrading, because when there's a breakdown automatic docking is abandoned. It was more rustic, but less attractive for a technician.

"Ficheur winced at this; he thought about it for a few weeks, and finally gave the go-ahead. I have to say that automatic docking in case of breakdown was included in the original specifications. I myself had suggested that we give up docking completely. You'd send someone. I'd said that in the long run there wouldn't be many breakdowns. But in the beginning there'd be too many breakdowns to send people to fix things by hand.

"That's always the question: Do you size things for the transitional phase, or for the permanent phase?

"You see, we really butted heads with the Lille authorities. In public, we were stubborn. On weekends, we talked things over more calmly . . .

"Finally, I let myself be convinced. We kept the docking, but not the CMD. In the long run, it's been a plus. We're trying to sell this 'plus' abroad, but it's

*For a long time, the only entirely automated systems that were really operational were at the Atlanta airport.

hard; you'd need to build it into the specifications, and that's not easy when you're dealing with other countries.

"I should say that there were only about fifteen people in charge. There weren't a lot of hierarchical levels. *Far fewer people* were involved than with Aramis. Broadly speaking, when Ficheur, Rullman, and I were in agreement, plus Notebart in the political arena, plus the Matra bosses, then that was enough.

"The atmosphere wasn't sectarian, and, more important, people *weren't looking for technological performance*. In Aramis, on the contrary, technological performance was an end in itself.

"The possible exception with VAL was the one-minute interval. Ficheur insisted absolutely on that. From time to time I'd say to him, '*Give me* a minute and a half, and I can simplify your whole thing.' 'There you go, Frèque, you're downgrading my whole system.' He stuck to his guns; I admit he was right. Finally we got there.

"The arguments sometimes got pretty lively. You heard everything: 'Greedy industrialist!' 'Profiteers!" 'Assholes!' But in the long run we reached an agreement.

"The problem with Aramis is that not enough people yelled at each other. Below a certain level, that's not good.

"You see, sometimes my ideas got rejected, other times I came out the winner; sometimes things got simplified, other times they got complicated. That proves it was a real debate, *a real negotiation*." [no. 41]

There are two models for studying innovations: the linear model and the whirlwind model. Or, if you prefer, the diffusion model and the translation model.

The two trajectories are quite different. In the first model, the initial idea emerges fully armed from the head of Zeus. Then, either because its brilliant inventor gives it a boost, or because it was endowed from the start with automatic and autonomous power, it sets out to spread across the world. But the world doesn't always take it in. Some groups, blinded by their petty interests or closed-minded when it comes to technological progress, are jealous of this fine idea. They downgrade it, pervert it, compromise it. Sometimes they put it to death. In certain miraculous cases, nevertheless, the idea survives and continues to go its way, a fragile little flame that burns in people's hearts. Finally, with the help of some courageous individuals who are open to technological progress, it ends up triumphing, at the price of a few adjustments, thus covering in shame those who hadn't known enough either to recognize it or welcome it. Such is the heroic narrative of technological innovation, a narrative of light and shadow in which the original object is complete and can only be degraded or maintained intact—allowing, of course, for a few minor adjustments. A religious narrative, naturally: a Protestant narrative, a Cathar narrative.

In the second model, the initial idea barely counts. It's a gadget, a whatchamacallit, a weakling at best, unreal in principle, ill-conceived from birth, constitutionally ineffective. A second difference stems from the first: the initial gadget is not endowed with autonomous power, nor is it boosted into the world by a brilliant inventor. It has no inertia. A third difference stems from the first two: the initial gadget moves only if it interests one group or another, and it is impossible to tell whether these groups have petty interests or broad ones, whether they are open or resolutely closed to technological progress. They are what they are, and they want what they want. Period. So how, under these circumstances, can the whatchamacallit interest anyone at all? By translating, as we know, from another mode and into another language, the interests of these groups. Hence the fourth difference: every time a new group becomes interested in the project, it transforms the project—a little, a lot, excessively, or not at all. In the translation model, there is *no transportation without transformation*—except in those miraculous cases where everybody is in total agreement about a project. Hence the fifth and last difference: after many recruitments, displacements, and transformations, the project, having *become* real, then manifests, perhaps, the characteristics of perfection, profitability, beauty, and efficiency that the diffusion model located in the starting point. A Catholic narrative. A narrative of incarnation.

There is something in the Aramis narrative that links it with the first model, and something in the VAL story that links it with the second. VAL starts small, and gradually becomes more complicated as it recruits local authorities who are all interested in the project as a way of advancing both their own affairs and the VAL line: Villeneuve-d'Asq's EPAL is developing the image of a new city along with VAL's cutting-edge technology; in a single breath, Professor Gabillard, a recruit himself, talks about hyperfrequency, the university, his own career, and VAL; Notebart, interested, pushes the Lille community, his own career, the prestige of the region, and also VAL; Ficheur, the contracting authority, excited about the project, is also pursuing his own all-too-short career, automated transportation systems, and VAL; and of course Frèque, Etienne, Lagardère, and Matra are advancing their own careers, their capabilities, their company, their stockholders' money, and VAL. All these interested people transform the project and put conditions on their interest: Notebart wants a network that obliges the nonreversibility of the early VAL to vanish; Ficheur holds stubbornly to his one-minute interval; Frèque doesn't want the CMD, which is making his life too complicated. What do they all do? They argue. They insult each other on occasion. In short, they negotiate and transform the project as often as they have to for it to end up holding its ground: the one-minute interval stays, the CMD goes, the network gets longer, the cars become nonreversible . . .

With Aramis, the general trend of the negotiation is quite different. The initial idea, an exciting one, of a point-to-point network served by a mobile unit with nonmaterial couplings is downgraded, but it is not renegotiated; it hesitates to locate itself somewhere, but it is not recombined from top to bottom. Aramis looks like a utopia, in the etymological sense—like an idea that has no place to land. "Where are we going to put this bloody mess?"

"I don't understand any more, Norbert," I confessed, increasingly uneasy. "Is it because Aramis is not negotiable that it isn't negotiated during that six-year period, or is it because it isn't negotiated energetically enough that it isn't negotiable and that it stays in its unfinished, hesitant state, stuck between being a mobile unit that is reaching perfection and a system that drifts with the wind?"

"Everything is negotiable."

"Not necessarily. If the very idea of Aramis is a take-it-or-leave-it affair, if it's all or nothing, it can be downgraded, modified, compromised without ever being renegotiated. That's what Girard implies: 'By giving up the point-to-point principle, they killed the project right then.' "

"But there can't be an intrinsic idea of Aramis; that would mean returning to a diffusion model, to the autonomy of technologies, to their irreversibility, their inhumanity. Everything we're doing in our research center runs counter to that idea. We might just as well say that Aramis has an inherent flaw."

"Well, why not?"

"I've told you a hundred times: because perfection is never inherent; it always comes at the end of the line. Talking about inherent flaws is the easy way out—it's a retrospective accusation. 'The thing doesn't work because it couldn't have worked.' It's a tautology. It's immoral; it's kicking a thing when it's down. It's the only crime our sociology can't forgive."

"Well, forgive me, but what if Aramis were in contradiction with your theories?"

"Impossible."

I felt like laughing, but when I saw that my mentor was serious, I bit my tongue and began to have doubts about his ability to solve the riddle of Aramis' death.

[INTERVIEW EXCERPTS]

In an urban research bureau run by the Ile-de-France Region, Mssrs. Grinevald and Lévy, are delighted to talk about Aramis.

"I'd like to reopen the question of technological feasibility, in relation to the question of profitability."

M. Grinevald:

"A simplified Aramis, downgraded, a little bigger, if it's significantly downgraded with five missions—*it might have worked. Intellectually, it's not inconceivable,* but we would have overinvested in supply because of the lack of flexibility and, in any event, it was very expensive.

"That's our president's point of view: it works, but it's expensive. He's convinced that it's a question of profitability, that it's like the Concorde. In my opinion, however, it's not because Aramis isn't profitable that it isn't in operation today; it's like an unprofitable Concorde, that's true, but *one that also functions at subsonic speeds!*

"Aramis is intellectually conceivable in a downgraded phase, but the nominal Aramis *is not* intellectually conceivable unless it is really outsized and completely ignores economic conditions and works as it would at a fair, without many people on board.

"Either the system has to be simplified, and then it's no longer a PRT, or else it has to be built as a PRT, and then transports nobody, or hardly anybody; there'd be a whole lot of empty cars. So in addition, it's *an unprofitable and passengerless subsonic Concorde!* . . . *[Loud laughter]*

M. Lévy:

"Yes, Aramis was put into intensive care; the thing was on full life support."

"So, for you, there's no mystery about it, because it was technologically infeasible?"

Grinevald:

"The problem, let me tell you, is that Aramis is a false invention, a false innovation. The PRT, from the beginning, was an *infeasible* idea from the operational standpoint." [no. 33]

"There you are, sir!" I gloated. "Wasn't I right? Congenital defect! I'm not the one who says so; it's the actor himself, as you put it. It's quite simply an inconceivable idea—that's what I've thought from the start, because I'm an engineer and you're not. It's perfectly self-evident that it's a false innovation."

"I'm telling you that's impossible. Grinevald and Lévy are bitter. You aren't born feasible or infeasible; you become so."

In the Aramis story, as in my mentor's obstinacy, there was a mystery that eluded me, a source of incomprehensible resistance. On the one hand, the engineers didn't want Aramis to be downgraded; they held to its essence come hell or high water—though they had progressively improved the essence of VAL. On the other hand Norbert, a perfectly preserved existentialist in the year 1988, was asserting that existence preceded essence; the absolute cynicism of his translation model looked to him like the only source of certainty and morality.

I am not yet among the powers that be. I am only a light breath, a feather drifting with the winds, a murmur in an engineer's ear, a wasp to be flicked impatiently away, an attractive idea that flits from seminar to colloquium to investigatory body to research report. "Aramis" is an argument, a story that grownup children tell themselves: "What if I were an automated car? . . ." It's an anecdote that moves from hand to hand and stirs up engineers; a rousing possibility. As a story, it is not yet cast in lead. It has not yet been replaced, in each of its words, each of its lines of blueprints, by a steel bar, by an aluminum plate, by a printed circuit. My story is told in words and drawings; it is not yet set in hard type. What the account book foresees is not yet accounted for, inscribed, engraved, burned in forever in the amorphous silicon. People stammer out my name. Nothing happens between two elements of Aramis that the engineers aren't obliged to relay through their own bodies. The motor breaks down, the onboard steering shakes and shatters, the automatic features are still heteromats overpopulated with people in blue and white smocks. Chase away the people and I return to an inert state. Bring the people back and I am aroused again, but my life belongs to the engineers who are pushing me, pulling me, repairing me, deciding about me, cursing me, steering me. No, Aramis is not yet among the powers that be. The prototype circulates in bits and pieces between the hands of humans; humans do not circulate between my sides. I am a great human anthill, a huge body in the process of composition and decomposition, depending. If men stop being interested in me, I don't even talk any more. The thing lies dismembered, in countless pieces dispersed among laboratories and workshops. Aramis, I, we, hesitate to exist. The thing hasn't become irreversible. The thing doesn't impose itself on anyone. The thing hasn't broken its ties to its creators.

4

INTERPHASE: THREE YEARS OF GRACE

"We must be closing in on our target by the sheer process of elimination," Norbert announced with a confidence that struck me as excessive. "Neither the very end nor the very beginning nor the Orly phase is to blame. And that whole horribly confused period from 1973 to 1980 can't be suspect either: when the engineers—yes, I do mean the engineers—do their summary report, whatever you may say to the contrary, just look at what they write!"

[DOCUMENT: RATP GENERAL REPORT ON THE END OF PHASE 3A, JANUARY 1981; EMPHASIS ADDED]

6. Technical Summary of Phase 3A

Despite a five-month delay in signing the agreement, and despite the many modifications required by the vehicle, the technological verdict on this phase is *very positive* for the following reasons:

—the track and its various components exist in a *quasi-definitive version;*

—the maintenance setup is operational and the automatic sequencing of tests has been demonstrated and implemented;

—the test runs of the car have shown that its various sub-components work properly and perform as anticipated.

In conclusion, the results observed during Phase 3A *do*

not invalidate any of the hypotheses proposed during the earlier phases.

However, the development of a sufficiently powerful motor, the reduction of the minimal turn radius to ten meters, and control of the cost of the electronic components require *progress or fine-tuning*. [p. 19]

. . . . .

8. General Conclusion

8.1. On the *technological level,* the results of the various tests lead to the conclusion that the completion of the center for technological experimentation can be undertaken in view of *homologation* of the Aramis system.

8.2. Comparing various possibilities led to the retention of the Petite Ceinture site in Paris, along the boulevard Victor, for the creation of the Center for Technological Experimentation, with an eye toward developing, on the basis of this initial infrastructure, *a first line for commercial experimentation* leading to the Exhibition Park at the Porte de Versailles; this line will be designed so as to allow for a significant extension all along the southern part of the Petite Ceinture.

8.3. The search for sites and the comparative studies undertaken elsewhere confirmed the hypothesis that Aramis might constitute an *interesting solution* for *filling in the gaps* in service in *certain* sectors of the greater Paris region, either by improving the feeder lines to the railroad network, or as an alternative to an extension of the metro, or by filling in the gaps in the rail network, or by creating a local network that either would be focused on a suburban center or would provide internal service for a major equipment zone, or else through a *combination* of these functions . . .

8.4. Given that, as in the earlier phases, the various results of the work undertaken in Phase 3A have *once again* been evaluated favorably, the RATP *supports the rapid completion* of the development of the Aramis system and thus considers that the next phase must now be inaugurated, leading to the construction of the Center for Technological Experimentation on the boulevard Victor. [p. 33]

"So for them, contrary to what Grinevald and Lévy say, everything is going along swimmingly. For the RATP, the decision is imminent. All that's left is to construct Aramis while refining the details that will allow it to go into production."

"Yes, Norbert, but if you look at the chart where we sum up the phases (the line shows where we are now—see?), practically nothing happens between this euphoric document and the July 1984 decision to create the CET [see Figure 1 in Chapter 1]. What could they have been up to for three and a half years? They were dragging their feet, weren't they?"

"That's true. It's odd, they want to get it done quickly . . ."

"And they wait forty-two months!"

"And in 1987 they finish it off rapidly. Poor Aramis."

"You said it! In three months, a quick thrust to the jugular . . ."

"It's an extraordinary case," Norbert continued. "The project is completely ready to go, according to the engineers, but it's also completely stalled. The whole thing must have played itself out in this interphase, right after the presidential election, during the Socialists' period of grace."

"Ah! I'm finally going to see the connection between technology and politics that you've been beating me over the head with for three months."

"If that's what you think, my friend, get ready for some surprises."

The more a technological project progresses, the more the role of technology decreases, in relative terms: such is the paradox of development.

As a project takes shape, there is an increase in the number, quality, and stature—always relative and changing—of the actors to be mobilized. Petit was just one highly placed official. Now ministers and presidents are involved. By moving from conceptual phases to preproduction phases, you move from saints to the God they serve. Since the project is becoming more and more costly, since it is agitating more and more people, since it is mobilizing more and more factories, since the nonhumans it has to line up are numbered in the thousands, since it is a matter no longer of plowing a beet field but of tearing up parts of southern Paris, actors capable of providing resources adequate to the new scale must henceforth be reckoned with. Ten times as many actors are now needed for the project, and they cannot be recruited one by one—one pipe smoker after another, one iron bar at a time. We have to move from those who represent small numbers to those who represent large numbers. In other words, the actors who are recruited and interested have to be spokespersons who already aggregate resources that are themelves multiplied: the region, the Paris mayor's office, the Left, the Right, France, industry, the country's balance of payments, exports. But the more the fate of the project is bound up with these new participants, the more room they take up, comparatively speaking. The only thing a technological project cannot do is implement itself without placing *itself* in a broader context. If it refuses to contextualize itself, it may remain technologically perfect, but unreal. Technological projects that remain purely technological are like moralists: their hands are clean, but they don't have hands.

This change of scale makes a project resemble an onion in which each layer of skin would be ten times larger than the one before, and it accounts for the impatience of the public. So: Is Aramis finished or not? Is it real or not? The answers to these questions depend not on the earlier stages but always on the latest one, which is also the most expensive, the most troublesome, the fussiest, the most complicated. Every two years the project bets its life on the red carpet, double or nothing! Ninety-eight percent of projects disappear in this game of roulette—so much for the fine folk who complain about irreversibility and the autonomy of technology. How the Matra and RATP engineers would like Aramis to be irreversible and autonomous! Would to heaven that the fine folk were right. The project engineers would light candles to Saint Ellul—never mind that he's Protestant—if it would help Aramis lose a little of its discouraging reversibility.

[INTERVIEW EXCERPTS]

At the Ministry of Transportation, a technical adviser is speaking:

"I have trouble seeing why you're doing this study; there's really no mystery here. May 1981—you do know what that was all about. And March 1986 as well.* The Left brings Aramis back to life and the Right kills it off.

"Each time, there was a lag of a year or two; that's normal, given the snail's pace at which they deal with dossiers. The technologists stubbornly kept on with their work. Not the politicians, who'd changed their minds. That happens a lot. Nothing can be done in this country without the shock waves provided by elections. The politicians have been completely inconsistent in dealing with this project."

"But it seems to me, since it's general and applies to all projects, that an innovation ought to be able to hold up against these changes in political personnel. VAL lived through two changes of administration; SACEM did, too, and so did the Poma-2000, the Rafale, and who knows what else."

"Yes, but Aramis was much more vulnerable to variations in the political environment; no one really wanted anything to do with it."

"So there are other ways to account for this vulnerability?" [no. 13]

"You don't seem to believe in the political explanation."

"No. That's all Big Politics. Wait till Wednesday; you'll see whether or not Big Politics explains anything."

[INTERVIEW EXCERPTS]

M. Piébeau, an economic adviser at the Transportation Ministry:

"What is completely left out of your question, Sir (although it does include interesting assumptions that I'm not going to dispute), is the economic dimension.

*François Mitterand, a (lukewarm) leftist, was elected president in 1981; Jacques Chirac, head of the Gaullist Party, was named prime minister in 1986, in the wake of an electoral defeat for the Left.

"You want to solve a mystery, do you? Fine. But there is no mystery as far as I'm concerned.

"Aramis works very well, but the economy hasn't kept up. Aramis began in a period of euphoria; moreover, a pretty rosy picture was painted of what the project would cost. Okay, but then the costs went way up and we went into a period of crisis. So Aramis no longer had priority. It's not much more complicated than that. It's a question of ebb and flow. Like tides. Aramis was floating along at high tide; then it washed up on the sand."

"Excuse me, but that strikes me as an unfortunate metaphor. In Brittany, if you go boating, you get poles ready, or a removable keel, so you can run aground safely. Whereas it seems to me the Aramis project didn't anticipate the reversal of the flow. It thinks it's still sailing in the Mediterranean, where, as everyone knows, there aren't any tides."

"You're playing with words. You can't prepare yourself for a major economic crisis—an oil crisis, for example."

"Sorry to lean so heavily on this point, but the decision to construct the CET was made in July 1984, ten years after the oil crisis."

"Profitability—you're not going to jettison that, are you? A project has to be profitable, at least a little bit."

"Again, forgive me, but all the economic reports during the entire fifteen years devoted to the Aramis project—all except the last one—were favorable to Aramis, highly so."

"But we all know what economic studies are worth."

"That's my point, that's exactly my point! So there are other reasons; there must be . . ." [no. 8]

"The economy, then?"

"I don't believe in the economy any more than in politics. The economic calculus—it's like that American joke: 'When my wife and I agree, my wife makes the decisions.' We'll have to look at the profitability calculations, but I have the impression they came after the decisions were made. In any case, in public transportation, according to what Liévin told me, what is profitable? With the compulsory transportation tax of 1 percent or even 1.5 percent on every salary, every year?* You could pay for Aramis. Strasbourg's VAL, the VAL planned for Toulouse, the Orly-Val that's in the works—are they profitable? The transportation tax is the envy of the world, it would appear. The Economy is much too big a thing to explain Aramis. We're not really going to drag Last-Minute-Determinations out of their beds and wheel them off to our metro. We'll see on Thursday."

[INTERVIEW EXCERPTS]

M. Maire, one of RATP's directors, is speaking in his plush office at the agency's headquarters:

"It seems to me that you're rather grasping at straws, if I may say so. You won't make any sense of this story if you don't talk about the Budget Office. The Budget Office rules France.

"It is—since you know the phrase now, and if you'll pardon the expression—in a state of intrinsic security. In a normal situation, you get Mr. *Nyet:* 'No money.' It takes enormous pressure to lift the ban and untie the purse strings.

"Okay, the Fiterman period was a nightmare from the Budget Office point of view. A strong and well-organized Transportation Minister—we'd never seen that before. I have to say that the guys in the Budget Office were getting worked over every day.

"Fiterman signed the CET the day before he resigned—maybe three days before, I don't remember exactly. It was when the Communist ministers left the administration [July 1987]. So what happened? The Budget Office did what every intrinsically failsafe system does: in the absence of pressure in the other direction, it froze everything.

"Aramis progressed against the Finance Ministry only *because of the completely exceptional circumstances of the post-1981 situation.* As soon as its

*In 1973 the mayors of Lyon and Marseille—Messrs. Pradelle and Defferre, respectively—transformed a temporary tax on salaries, intended to finance their subway systems, into a proposed law that was welcomed enthusiastically by local officials. A mayor may levy 1 percent (and even 1.5 percent, when heavy investments are involved) on all the salaries in his district to finance public transportation. This guaranteed manna, whose flow is tightly controlled by the local elected officials, makes it possible to ensure relatively high salaries for positions in public transportation and allows for very significant investments with borderline "profitability"—if this word means anything at all.

supporters lost their fighting spirit, it was all over. In this business the Budget Office is everything."

"But all expensive projects are subject to the same pressures by the Budget Office. This doesn't explain away Aramis."

"It's true that in aeronautics they're better organized. They have submarines, you might say, in the Budget Office *[Laughter]*. Over here in ground transportation, our relations aren't as good." [no. 22]

M. Gontran, researcher with the Institute for Transportation Research:

"Besides, there's a special problem with ground transportation. Its relations with the Budget Office aren't cozy—they're not like those of civil aeronautics, for example. Those guys have got Budget Office personnel assigned to them. It's like the Trojan Horse: they explain, they make things clear. But with the DTT it doesn't work so well.

"Even for the SK, they had to tell the Budget Office sob stories to get three million out of them, which is minimal. The Budget Office made a huge fuss over those millions, even though the Japanese bought the system!

"I have to say that this lack of confidence is due to failures, to the Bertin business. There are lots of skeletons in the closets." [no. 42]

"Now that one really has me convinced. The Budget Office is crucial," I said enthusiastically.

"But it's still too big, my friend. Why not Parisian centralization, Napoleonic France? Why not technocracy, while you're at it? Everybody talks about the Budget-Office-which-blocks-all-decisions-to-innovate. Forty Americans have written theses on the topic. No matter what the subject, the Budget Office is the obstacle. France has been carrying on for four hundred years, but Colbertism, Napoleonic centralization, and the Budget Office, according to our distinguished analysts, prevent any change on principle. 'Plus ça change, plus c'est pareil,'" he added with an American accent. "It's not reasonable. It's crude sociology."

"And what kind of sociology are we doing?"

"Refined, Mister Young Engineer, refined sociology which applies to a single case, to Aramis and only Aramis. I'm not looking for anything else. A single explanation, for a single, unique case; then we'll trash it."

A "trashable explanation"! The exact opposite of what I'd been taught about the universal laws of Newton and Einstein.

"So we're waiting for Friday?" I went on, without calling attention to Norbert's epistemological aberrations.

"Exactly. No, Monday. Friday's meeting has been canceled."

[INTERVIEW EXCERPTS]

M. Pierre, a Research Ministry official in charge of transportation studies:

"You have to put yourself back in that atmosphere. In 1981, in all the ministries, you had an influx of people who'd been out of power for twenty-five years. In research—which had been a stepchild for ten years, remember—there was Saunier-Seïté, the researchers' nightmare. Then Jean-Pierre Chevènement came along, determined to make the first Ministry of Research and Technology the big project of Mitterand's first term, and to endow it for the first time with a single, coherent budget—which he wanted to double. The little world of ground transportation didn't escape the mobilization.* And in particular, at the Transportation Ministry, Fiterman—one of four Communist ministers—took over; you'll remember how their entrance into the administration 'shook up the Western World.' Fiterman wanted to take a relatively weak technological ministry and turn it into a major one; he wanted to bring infrastructures—roads—and public transportation together for the first time. In August 1981 another Communist, Claude Quin, was named president of the RATP. He wanted to make that outfit a showroom both for social relations and for the technological quality of service. In January 1982 the managing director of the RATP was replaced by the former director of the railroad, Girard. *The new political context?* They were it, and they really did represent a significant change, or at least the desire for a significant change.

"As for the handful of RATP engineers, this change was rather disturbing. The new director was known to be skeptical about Aramis, if not downright hostile. People were sure he was going to kill the project. For Matra, a company that could be nationalized, the success of the Left and the installation of two Communists in positions that were key for the metro—these develop-

*In France, roughly a thousand researchers specialize in transportation: some 300 are employed by the major automobile manufacturers; another 300 work for major enterprises, the SNCF, and the RATP; and the rest are scattered among public research organizations.

ments boded no good. Aramis was at a standstill, in a position of extreme weakness, on the eve of an enormous request for funding for the construction of the CET. The only element that augured well for the researchers in ground transportation was the Left's stated desire to develop industrial research and to bolster the image of France as a nation on the cutting edge of technology. You have to put yourself back in the atmosphere of the times. We hadn't seen anything like it since De Gaulle's early days."

"Finally, straightforward politics," I said delightedly. "We've got them all, the Commies, the pinkos, Big Politics. It wasn't so complicated after all. Contrary to what you were saying, Norbert, it wasn't worth the trouble to go into all the technical details. All we needed to do was look at the overall context. He's right: it's a question of atmosphere."

"Atmosphere, atmosphere," my mentor repeated drily.

A technological project is not *in* a context; it gives itself a context, or sometimes does not give itself one.

What is required is not to "replace projects in their context," as the foolish expression goes, but to study the way the project is contextualized or decontextualized. To do that, the rigid, stuffy word "context" has to be replaced by the supple, friendly word *"network."* The big explanations in terms of politics, economics, organization, and technology always turn up, without fail: "It's politically unacceptable." "It isn't profitable." "Society isn't ready for it." "It's inefficient." These explanations are always used precisely because they can't be worn out. They're not designed to explain—if that were the case, they would have to wear out in contact with the hard, contorted circumstances. Rather, they're intended to move from hand to hand and to serve, like the weasel in the children's game, to get rid of the problem by designating the one who failed to pay attention and got caught with the ring in his hand.

Il court, il court le furet	The weasel is running,
Le furet d' la politique,	The political weasel;
Il court, il court le furet	The weasel is running,
Le furet d' la technique.	The technological weasel.

Il est passé par ici	He came this way;
Il repassera par là	He'll go back that way.
Il court, il court le furet	The weasel is running,
Le furet d' l'économie.	The economic weasel.

This is what children sing as their moist hands polish the ring that is passed around on a string.

To get rid of one's own responsibility, the big explanations are useful; but as soon as one stops trying to blame someone else, these big explanations have to be replaced by little networks. Who decides that Aramis must be influenced by the change of administration in 1981? Four or five people, all identifiable and interviewable. Who decides that aeronautic projects don't have to be subject to the vagaries of changes in administration? Ten people, some of whom, it is said, are camping out in the Budget Office as "submarines" or "Trojan horses." Who decides about Aramis' economic profitability? Eight people, all identifiable and interviewable. Who decides that the economic calculations that prove Aramis' profitability are pure fictions? Again, four or five people, the same ones or others, equally identifiable and interviewable. Who passes judgment on Aramis' technological feasibility? Three people, maybe four. Who passes judgment on Aramis' technological infeasibility? Fifteen people or so; they're harder to pinpoint, but their tongues loosen after a few hours of conversation. It's clear: Aramis is not in an overall context that has to be taken into account. To study Aramis after 1981, we have to add to the filaments of its network a small number of people representing other interests and other goals: elected officials, Budget Office authorities, economists, evaluators, certain members of the Conseil Supérieur des Ponts.

The few elected officials recruited by the project certainly don't count as Politics; the economists who calculate profit margins don't constitute Economics; the handful of engineers who evaluate Aramis' technological refinement certainly don't equate with Technology. The impression of a context that surrounds the project comes from the fact that one forgets to count the handful of mediators who speak in the name of money, Official Bodies, chips, or voters. Once we add the spokespersons back in, everything clears up: the network is extended, but its nature doesn't change. We've gone from a network to a network and a half.

"I don't understand why you don't want to give any importance to politics, to the context. You're a sociologist, after all."

"Precisely because we're in the process of making a sociological discovery. Because we've had the luck to come across a unique case that's never been described by anyone. In 1981 Aramis was at a standstill. It had no inertia. It could disappear without shocking anyone, or veer to the right, or . . .

"Veer to the left."

"Yes, but without any impetus. They really went looking for it; they picked it up. That's what's unique. The zero-degree inertia. Mr. Britten even showed me a 1986 report that compares all the Personal Rapid Transit systems in the world. Do you know what the Canadians said in their report?"

[DOCUMENT: FROM R. M. RENFREW, M. L. DRISCOLL, AND K. ROSE, *PEOPLE MOVER MARKET REVIEW* (CANADIAN INSTITUTE OF GUIDED GROUND TRANSPORT, 1988)]

The concept of the people mover as a technology response to mobility problems gathered momentum in the 1960s as a consequence of broad administrative support—and in very rare cases, support by transit operators. It received unprecedented industrial interest when the aerospace industry in Western Europe and the U.S. came to the conclusion that innovative transit represented a significant new business area to compensate for declining aerospace opportunities. [p. 1].

The bubble burst rapidly and catastrophically. From 1975 to 1980, there was a retrenchment to completion of commitments made earlier—frequently truncated—and a modest technology program . . . until the core groups disappeared by attrition and eventual termination by management. Long before the cardiac arrest of the industry became final, the transit operators in Germany, the United Kingdom and the United States had ceased to follow the people mover development at all. Most of them became bitterly opposed to the diversion of public funds which they felt should have been employed in building reliable, predictable, conventional transit systems. [p. 4].

"So you see? In 1980 all it took was to do nothing at all to kill Aramis, since everywhere in the world—Germany, Japan, the United States—PRTs were in a state of 'cardiac arrest,' as they say. Moreover (and this confirms our hypothesis about the innocence of the earlier phase), after Orly all PRT projects had the same defects: hesitation, absence of normalization, cost overruns, delays, total chaos. So there's nothing unusual in this period. The project is normal."

"Do they mention Aramis in that report?

"Yes, of course. On page 65 they say—speaking of the year 1986—that Aramis is 'the only credible PRT system under development at this time.' They say that it's crucial for the future of PRTs because of the rotary motor, the platoon configuration, and the automated self-diagnostic mechanisms, but that it costs a small fortune."

[INTERVIEW EXCERPTS]

M. Girard, speaking in a temporary office:

"When I was appointed *[as head of the RATP]* in 1982—I'll be honest with you—when I opened the file, I wanted to close it right then. 'Enough fooling around with that stuff; we can place our bets more effectively than that.'

"Lagardère probably had the same thought. 'Every day brings its own troubles. VAL is bad enough—why run after mirages?' The head of the RATP, the head of Matra, in their heart of hearts, *were determined to kill Aramis.*

"I myself *became a convert,* for two reasons. First, I went back upstream, as it were, back to the somewhat utopian thinking of the 1960s. It's still a current issue, too: we need *something like* cars that join together, trains that split apart. A bit maliciously, I even thought it would work for the existing network, especially the RER. Aside from a few very crowded transfer stations, the RER amounts to a lot of long, empty cars that we drag around the countryside. Couplings and uncouplings are terrible problems. Cars that come together smoothly might well come in handy. I said to myself, 'Hey, *in five or ten years,* we'll need a system like that for the branch lines.'

"The second train of thought that accounts for my 'conversion,' if you will, was the 1989 World's Fair. Every exhibit presupposes a new form of transportation. Within the gamut of what was being proposed, Aramis was really innovative: France could really present something that would symbolize French technology at the end of the 1980s. That's what made me change my mind." [no. 18]

The work of contextualizing makes the connection between a context and a project completely unforeseeable.

The history of technology—like history, period, "big" history, full of sound and fury—is at the mercy of a reversal, an overturning, a bifurcation. The new company head opposed Aramis when he was with the Rail Division. He opens up the file. What influence will he decide to exercise? As a builder of major metro systems, will his cultural hostility to this toylike jerry-built gimmick come to the fore? Or will his president's fondness for a major technological project capable of mobilizing the energies and enthusiasm of engineers win out? A moment of uncertainty. A crossroads. *Kairos.* The word "conversion" has to be taken seriously, especially when it is uttered by an engineer who is also a theologian. The company head meets his road to Damascus. He was against Aramis. Now he is for it.

In fact, every element of the aforementioned context decides, or not, to be a conductor of influence, a semiconductor, a multiconductor, or an insulator. Transforming the context into a certain number of people who represent interests and who all want to achieve the goals of those they represent thus does not suffice to enable one to decide whether or not they will have any impact on Aramis, still less to calculate in advance what the impact will be. In a given context, the same projects do or do not feel an impact; a single context can bring about contrary effects. Hence the idiocy of the notion of "preestablished context." The people are missing; the work of contextualization is missing. The context is not the spirit of the times which would penetrate all things equally. Every context is composed of individuals who do or do not decide to connect the fate of a project with the fate of the small or large ambitions they represent. The new people that elections bring to power may decide to make Aramis one of the great ambitions of the technological Left; but they may also decide that Aramis is one of the many projects that the Right is dragging in its wake, projects that devour public funds vampire-fashion and that have to be finished off, like the aerotrain, with a solid stake through the heart.

Aramis can be contextualized, but it can also be *decontextualized.* This freedom to maneuver is all the greater in that the project finds itself at a standstill, and in that all the PRTs in the world have been dismantled or suspended. To bury Aramis, it is enough, literally, to do nothing. The new people may decide that Aramis belongs neither to the Right nor to the Left, and thus may abandon it to the limbo it has kept on entering and leaving since the Orly phase. They may also decide nothing at all. They may never even hear a word about Aramis. How do you account, then, all you contextualists, for the fact that in May 1981 Aramis could have gone completely unnoticed, could have been on the Left, could have shifted to the Right, could have transformed itself entirely to become a mini-VAL, could have called itself Athos or d'Artagnan? "It's no accident," you say? No, indeed, nothing happens by accident; but nothing happens by context either.

Clearly I had judged him too quickly; perhaps my professor was less incompetent than I'd thought. What had happened was that the label "sociologist" had been leading me astray from the start. He never used the social context as a starting point. On the contrary, the social context was what he was driving at: he wanted to explain it. I should mention that our interlocutor, M. Girard, was so precise, so frank, so cultivated, that in fact he did the sociology all over again in our stead, just as Norbert had predicted.

[INTERVIEW EXCERPTS]

M. Girard again:

"I presented the idea to Lagardère. Lagardère *let himself be drawn in,* reluctantly. Of course, he was interested by the 1989 showcase *[the World's Fair],* but he lacked his usual enthusiasm.

"That was when I was betrayed by the World's Fair technicians and experts. The RATP people told me, 'We won't be able to bring it off within the allotted time frame unless we can use safe, tested technologies—that is, no nonmaterial couplings on the cars.'

"In other words, it was a *denatured* project. What's more, there were people connected with the fair who wanted an ironclad guarantee that we would connect the two sites, Balard and Bercy: 'We're ready to do it only if there's a full technological guarantee.'

"What was I supposed to do? I took a chance on a dubious operation. I tried out the simplified solution; that way I could reassure the experts, and if it worked I could *reintroduce* nonmaterial couplings. I was sure I would get somewhere.

"Of course, it wasn't elegant—it was a compromise." [no. 18]

"Watch out now, it's getting tricky; we have to note everything, because the possible Aramises are about to multiply. Girard is ready to kill:

"1. a gadget-Aramis that has provided enough amusement already.

"On his side, according to him, he has:

"2. an Aramis that is preventing Lagardère's VAL from progressing ('every day brings its own troubles').

"But he's ready to defend:

"3. an Aramis that is composed of cars capable of linking up in five or ten years, and that will allow the RER or the metro to handle branch lines.

"He's also ready to defend:

"4. an Aramis that will serve as a French showcase for the 1989 World's Fair.

"Let's keep tallying them up, all these Aramises, because their lives are at stake:

"5. an Aramis with rendezvous capability for the World's Fair that doesn't interest Matra very much;

"6. an Aramis with rendezvous capability for the World's Fair that is infeasible in the time frame allowed, according to the technicians;

"7. an Aramis for the World's Fair that is technologically tried and true and that doesn't have rendezvous capability."

"Number 7 is incompatible with number 3, since it no longer has rendezvous capability, and it's almost incompatible with number 4, since it isn't really new."

"Yes, but because it's denatured, it has the support of the World's Fair technicians. That gives us:

"8. A denatured, compromised Aramis, reluctantly supported by Matra, feasible in the allotted time frame according to the technicians, with no rendezvous, for the World's Fair; and:

"9. the same, with the later addition of nonmaterial couplings, which takes us back to number 3 but not to number 4.

"It's true that it's not elegant."

"Seeing as how you're always in favor of sociotechnological compromises, Norbert, you must be happy."

[INTERVIEW EXCERPTS]

The same M. Girard, as direct and precise as ever:

"Then Aramis *became* a political symbol associated with the image of the World's Fair and with the left-wing administration supported by Quin, the *[RATP]* president and by Fiterman, the minister. Quin liked Aramis because of its research aspect, because it was good for France, because it was the flagship of RATP research; and it became almost a *political slogan of modernization* . . .

"If I'd been reasonable, I would have stopped it the day the World's Fair ended [in June 1983]. But Aramis was so *promising,* the administration had *invested* so much at the level of discourse, that I let things go on. I should have done something, but then I went back to dreaming, back to my RER, and I told myself, 'Maybe it's not so stupid,' *knowing perfectly well* that it wasn't very satisfactory.

"That's why I was interested in Montpellier— because the city was motivated, the traffic wasn't too heavy, and you could have a dense network, one with multiple origins and destinations. So we decided on the CET, but you must realize that the project always had that *congenital defect.* It wasn't *supported* by the RATP—it was *politically vulnerable."* [no. 18]

"One more congenital defect," I exclaimed enthusiastically, "and even two, but this time not on the technological side: on the political side, on the cultural side."

"Let's keep on counting the translations. The more complicated it gets, the more we have to stick to our little accounting processes:

"10. It's a technologically modern, political flagship of the Left. What Aramis means now is: 'We're modernizing the RATP.'"

"But what does it do technologically, this last Aramis?"

"We don't know what it does, but we do know that the president and the minister love it, as long as it's complicated enough to be a symbol of modernization."

"Ah, if there's love, then you're going to be happy, Norbert."

"Yes, because with my little chart I can deploy the loves and hates that fluctuate according to the various shapes of objects. Look, we have more:

"11. Aramis without the World's Fair no longer interests the company head, who again wants to kill it.

But:

"12. Aramis is loved by the president and the entire Left; they don't want to kill it.

"Hence an Aramis:

"13. with nonmaterial couplings for the RER, without the World's Fair, which is not stupid but not satisfactory;

"Followed by:

"14. an Aramis in Montpellier that the company head likes a lot, but that probably nobody else does, except for the mayor;

"and finally:

"15. the Aramis that is the CET in Paris, a stepchild of the RATP, with a bit of support from a smattering of people.

"Look at that plot, my young friend. If it were a play by Corneille, people would call it a miracle; they'd admire the violence of the passions, the intensity of the reversals. Yet we're dealing with automated subway systems and technocrats. This is the real literature of our day."

"Too bad the company director didn't kill off Aramis in 1983—we'd at least know what it died of."

"Yes, we've made a lot of progress. The mystery is not its death in 1987 or 1982, but why it came back to life between 1982 and 1987. Now that it's loved by so many people, we're going to have to move from the little guys to the big guys. Ready to take on the stratosphere, Mister Young Engineer? You must admit you're learning more here than in your classrooms, where you solve equations the professor already knows how to do. Do they talk about love, at least, in your school? They never teach technology in engineering schools, if they don't teach you to follow a project from the smallest cubbyholes up to the loftiest spheres. Our laboratory is Paris and its antechambers."

"Then Paris is ours!" I exclaimed in a moment of excitement for which the reader, aware of my youth, will surely pardon me.

Technological projects become reversible or irreversible in relation to the work of contextualization . . .

Aramis, delicately balanced up to this point, has become so weak that a puff of air could wipe it out: zero-degree inertia, maximum reversibility. Aramis, now linked to Big Politics, becomes "promising"—"so promising" that the company head can no longer stop it, he claims. Maximum irreversibility: no one, despite the desires of the two chief protagonists, can kill it. The type of irreversibility changes in just a few months—and will change again, several times, proving to what extent projects are of course *reversible*. This is because technological projects can be tied into different contexts and can thus become promising or needy, depending on circumstances. They are set into context by the spokesperson, as something is set to music or "orchestrated." Aramis in 1981 nearly becomes a pile of paper covered over by the drab surfaces of closed files. It nearly becomes one of those thousands of projects that slumber in engineers' drawers and studies. The company head could make it unreal. He has his finger on the button, ready to send it back to the void. But then something happens: the work of contextualization starts up and is so successful, so sprightly, that Aramis finds itself solidly ensconced on the Left. After getting Petit, Bardet, and Lévy enthused, all of a sudden it excites the Communist president. It has become a political slogan, a reference in so many speeches and in so many newspaper articles that it has a life of its own; it can't be stopped. For a project that was an amusing dossier that needed to be closed in a hurry, it's not doing badly at all.

Contextualization is fabricated and negotiated like everything else: by tying bigger and bigger pots and pans, and more and more of them, to the project's tail. When it stirs, it's going to stir up all of France. It makes enough commotion to wake up a minister. But the pots and pans still have to be found; they have to be tied on tight, and the beast has to be made to move. A lot of work that the notion of "context-given-in-advance" haughtily refuses to acknowledge. This neglect is all the more damaging in that such work can be undone. When Aramis dies, in 1987, no minister will stir. All the pots and pans will have been removed. Those who were counting on the irreversibility of the context to keep their technologies alive wake up in the cemetery.

[INTERVIEW EXCERPTS]

M. Girard is speaking once again:

"How do you explain the paradoxical fact that the CET started up again just when everybody finally was beginning to have doubts?"

"First, the backbone of the thing is the southern part of the Petite Ceinture, with its branch lines. There we developed a project that corresponded, *like it or not, to a certain number* of objectives. That project *made it clear,* perhaps, that extensions of the metro were financial crimes.

"For example, there was a plan to extend Line 4 toward the south; that would have completely downgraded the line. The Porte d'Orléans terminus is remarkable; it does its job in 80 seconds. We would have really messed things up on the operational level and spent colossal sums if they'd insisted on extending the lines. The RATP was burdened with a series of projects involving the extension of lines—projects that it didn't know how to get rid of. Aramis was *an alibi.* Everybody got excited about it; we spent three or four million francs for a *mediocre interest,* but in any event *no one said another thing* about extensions."

"Could you go back over the list of project supporters in 1983–84? This is the period I'm having the most trouble grasping."

"Okay, there was the Mauroy administration, which strongly supported it. It was supported by Quin. It was supported by me—I was very happy about it, although skeptical; I *would be able to nip in the bud* the line-extension projects that I really didn't like. Then there was Orly-Rungis *[predecessor of*

Orly-Val], and then Montpellier. We told ourselves: 'All these projects *are not stupid.* They're worth it. In any case, the technological development is *interesting,* and there's a *possibility* of industrial development' . . .

"The end didn't surprise me; all it took was a finance minister . . . It was a *colossus with feet of clay.* Meanwhile, all the supports for the colossus had disappeared. I should point out that there weren't any more suburban mayors clamoring for lines. The Paris mayor's office, for its part, remained *interested.* Chirac wrote to the RATP asking that something be done on the Petite Ceinture. Since the success of the Lille project in 1977–78, Matra *believed in nothing* but VAL, and it was all over for everything else. Aramis was just 'to see what would happen.'

"It hardly matters who provided the last straw that did the system in; that's a proximate cause. In any case, the point is that *a last straw was all it took.* It doesn't matter who killed the project. I don't actually know the proximate cause."

"But the remote cause—do you know that?"

"Yes, of course. You know, when I understood that Aramis had been terminated, I wasn't surprised; *for me, it was built into the nature of things.*"[no. 18]

"Why do we do the sociology and history of technology," asked my mentor Norbert, with tears in his eyes, "when the people we interview are such good sociologists, such good historians? There's nothing to add. It's all there. 'Built into the nature *of things:* there you have it—technology! Insert, engrave, *inscribe* things within, inside, right in the middle, of nature and they flow on their own, they flow from the source, they become automatic. Give me the remote causes—let's go back to the mainsprings of the tragedy—give me Matra, the Communists, the Right, the Left, the mayor of Paris, the traitorous technicians; let's put them on stage in 1984 . . . and in 1987, here comes the death blow. An implacable clockwork is operating before our very eyes. And it's he, the company head, who inscribes, who engraves, these things in nature. He himself machine-tools the *fatum* that is going to bring the plot to its conclusion with no surprises; he's the *deus ex machina,* the god of machines. Enshrine the interviews and shut up—that's the only role for a good sociologist."

Without letting myself get as worked up as he was, I went on enumerating Aramises, as he'd taught me:

16. Aramis on the Petite Ceinture with a few branch lines—it's interesting, in spite of everything;

17. Aramis makes it possible to nip in the bud the projects for metro-line extensions that the mayors wanted, and so it interests the RATP even if it doesn't get built;

18. Aramis with nonmaterial couplings no longer interests Matra, after VAL's success; it no longer interests the mayors either, but it still interests the technicians a little, and it interests Montpellier a lot, and also the Mauroy administration;

19. Something, but not necessarily Aramis on the Petite Ceinture, still interests the mayor of Paris;

20. Aramis is not loved by the finance minister.

"There's nothing more to be said," Norbert announced, rather irritated by my academic approach to the issue. "They're doing the sociology for us, that's all."

"If they're such good sociologists," I replied, bewildered, "if they do such good sociotechnological analyses, why not say so in 1984? Why not write it in the reports? Well, I want to point out that we find no trace of these refinements in the documents we've read. Not a word of doubt in the minutes of the development committee.* We find only constant praise of Aramis, the eighth wonder of the world. Neither visible compromise nor negotiation."

"Say, that's true! We'll make something of you yet, if the goblins don't get you . . ." It was the first nice thing he'd said to me since I started my apprenticeship.

That was the happiest point in the investigation. I now respected

*To follow up on major projects, the ministries can delegate the task of scientific oversight to a development committee that brings together all parties involved and that is charged with informing the public authorities about the evolution of the relationships between the contracting authority and the contractor. In practice, it functions too often like a rubber stamp.

Norbert's sociology, he respected my abilities, and each new interview brought us crucial information about this astonishing interphase.

[INTERVIEW EXCERPTS]

The RATP's M. Gueguen reminisces:

"In 1983 we made a presentation to the board. 'If this is how you're showing it to me,' declared the board president *[M. Girard],* 'I can only conclude that we mustn't do it. Show me some other way.'

"It wasn't really that there were problems, but there were a number of technological risks, and putting them all together made the project really marginal.

"You have to understand that the project was absolutely state-of-the-art in all respects.

"We all sat down around a table. Somebody said, 'Aramis is new all around; it's the wrong sort of step, just what mustn't be done.' At the next development committee meeting, *things were sorted out differently.* 'It's fine, no problems.' The risks just had to be turned into certainties. The risks had to be wiped out; then they said, 'Let's go.' The red light changed to a blinking signal.

"Girard was convinced, but obviously he wanted the thing to be well presented. The technologists couldn't stand up to the politicians. The current line is that 'technologically, it's a success'—but that's not how *we* put it. You can have a look at it, but don't copy it." [no. 2, p. 13]

[Fumbling through his documents, he gets the original handwritten minutes.]

[DOCUMENT]

'January 27, 1983: M. Girard is struck by the large number of obstacles mentioned during this presentation. It seems to him that everything has been done to make it impossible to pursue the Aramis project. However, this system strikes him as having considerable potential value, and it would be too bad if unavoidable delays forced him to give up the project . . . Consequently, he asks the head of technological services to prepare an attractive dossier for February 1, 1983, one that would justify a positive decision.'

"Hey, this is really heating up," I said, smacking my lips.

"I already told you, innovation studies are like detective stories."

"And now," he went on, "let's go see whether M. Girard's splendid analysis turns out to be confirmed on the other side, at Matra."

[INTERVIEW EXCERPTS]

At Matra, M. Etienne displays the same frankness, the same precision, the same knowledgeability:

"There came Fiterman and Quin, and Girard along with them, all new guys.

"Among their ambitions, one thing had struck Quin: the lack of a major research project at the RATP. He wouldn't have minded being able to tell his minister, 'We've made a big effort in urban transportation.' That couldn't leave the Communists cold. It's good for the workers, urban transportation.

"It wasn't stupid. Fiterman put up a good fight. It's too bad they didn't back the right horse. Also, Fiterman the Communist was all excited about working with Lagardère *[the capitalist]*.

"There were conflicts with the RATP right away; not with the politicians, but with Girard. He didn't believe in it at all, at first, but he couldn't be against everything; and besides, since he really likes to be in charge of things, he took this one over, made it his baby.

"In addition, he knew perfectly well that the job of metro driver is a hard one, that the way drivers feel is a delicate issue. You couldn't say that people were thrilled about losing a job that was sometimes passed on from father to son.

"We proposed an Araval to Girard. He told us, 'You do the nominal Aramis or we aren't interested.' Girard went to bat for it. I was convinced that the Petite Ceinture was viable; I wanted to take the quickest way, and I didn't want to see my company lose money. Araval was less complex, and it fit very well, with its small size; it was less cumbersome than the tramway. Girard would have none of it; he wanted the nominal Aramis.

"It's true—at that point I perhaps lost my nerve. *I didn't have the guts to say, 'No, Araval is better than Aramis.'*"

"But again, why did the RATP plunge into the most complicated project?"

"Girard didn't trust me. He didn't want me to foist off a simplified system on him. I never should have called it Araval. The name was a kiss of death. I should have called it Athos or Porthos!

"In fact, *I should have spent a couple of hours with Quin behind Girard's back, to convince him it would never work.*" [no. 21]

"Extraordinary! Compare this with the previous interview. Girard says, '*If I'd been reasonable, I would have stopped it the day the World's Fair ended.* But Aramis was so *promising,* the administration had *invested* so much at the level of discourse, that I let it go on.'"

"So they see eye to eye."

"Yes, they agree, but on a misunderstanding! In twenty years as a sociologist, I've never seen a thing like that. The two most important decisionmakers both think that they shouldn't have made the decision they made—that they should have been more courageous. Magnificent. Truly magnificent."

[INTERVIEW EXCERPTS]

Still in M. Etienne's office at Matra:

"As for Lagardère, he said to himself, 'Here I am in Notebart's position, with the local locomotive that's going to pull Aramis along for me.'

"He looked me straight in the eye. 'Can you do it?' I said, 'It's complicated, but we'll do it.' Even though I thought it was unnecessarily complex.

"Obviously, then there was the World's Fair; we thought that would help us, and we were *pushed along a path* that we never were able to leave.

"The paradox is that when there was no more World's Fair, Girard said, 'You see, simplifying wasn't worth it, since there's no rush' . . .

"Girard always leaned toward *the complex solution.* 'Real *high-tech,* M. Quin; not the semi-high-tech that Matra has the nerve to propose to you.' That's what Girard was saying.

"For my part, I knew perfectly well that things had to get more *complicated as they went along.* That's what Papa Dassault did. What do you make of a thing that goes "Daddy! Mommy!" right away?

"If I've understood correctly, the quickest way to get to a real-life product was the complicated way?"

"Yes, *the shortest route was still a long one* because it went by way of

Girard, who didn't trust us or our efforts to simplify. Yes, we'd accepted that." [no. 21]

"You can see what that guy loves and hates. Look, let's draw a line that connects all the variations in feeling to each variant of Aramis: what do we come up with? The *Carte de Tendre,* the *Map of the Land of Tenderness**—a new one for the century, ours, a map that has been neglected by novelists and the sniveling humanists. Yes, Aramis is loved, Aramis is hated. It all depends on its changing forms. Here's the Peak of Conversion of the company head; here's the Swamp of Cold Feelings; further on, the Temple of Enthusiasm of the World's Fair. Then you find Love's Trickery—ah! if only we'd called it Porthos!—and the Go-Between's Hut, and the Budget Office Cave, and the Pit of Disappointments, deeper than the Pacific Ocean . . ."

"And you get to Marriage in the end?"

"Yes, but a marriage of reason, a shotgun wedding."

"But since it concerns love of technology, Norbert, we should call it the *Carte de Dur*—the *Map of the Land of Hardness.*"

"Hey, that's great! Come here, let me give you a hug! A real love story. Yes, that's it!" exclaimed my mentor. "They fell in love with Aramis! If you're actually going to love technology, you have to give up sentimental slop, novels sprinkled with rose water. All these stories of efficient, profitable, optimal, functional technologies—it's the 'Harlequin Romances!' It's 'The Two of Us!' You get paid by the word, and you sell them by the truckload. It's disgusting. The two of us, my friend, we're telling real love stories; we're not naïve romantics."

The shortest path between a technological product and its completion may be the crookedest one.

*This map originally appeared in Madeleine de Scudéry's allegorical novel *Clélie* (1654). Using the conventions of what was then the new science of cartography, the map depicted love affairs as so many trajectories through an imaginary space, which Scudéry called the "Royaume de Tendre."

The rationality of technologies writes like divine Providence, straight across curved lines, crossbars, forks in the road. The geometry of the feasible and the infeasible, the complex and the simple, the real and the unreal, is as miraculous in technology as in theology. In fact, the trajectory of a project depends not on the context but on the people who do the work of contextualizing. Araval, a small-sized VAL, stems from a sociotechnological compromise that would simplify the Aramis project. Yes, but it would no longer interest the company head. Aramis for the World's Fair is a revolutionary transportation system that everybody is excited about. Yes, but the technologists are "traitors" and say "you have to have material couplings" in order to make the eighth wonder of the world feasible. "Not feasible," "not interesting," "not lovable." All these terms are negotiable. The result is a hybrid? Yes, an Aramis that is complicated enough to please the company head, who is interested in it because it's complicated and not very feasible, but also because it gets people used to automation, because it "nips line extension projects in the bud," because it offers a good image of high-tech research—and in any case it's already too late to decide, because the project has become promising. There's no stopping a modernizing political slogan.

As for the manufacturer, he loved Aramis. But he said he had made a mistake—still another past conditional, another "I should have." For his hybrid he invented a word that is itself a hybrid, based on the rival words "Aramis" and "Val," which the company head wanted to keep as distinct as possible. Ah! If only he had put "Porthos" on his rack! Our manufacturer was not a good Scrabble player. What's more, he didn't dare bypass the hierarchy to speak about his projects directly with the RATP president, the only one who could have overruled his company head.

Still another twist in the road? Yes, another short cut—a shorter, more meandering, more undulating route. Moreover, when his president took him aside in private, he said, trembling a bit: "It's doable." What beautiful, admirable symmetry! While the RATP technicians were rewriting the project of the complicated Aramis, which they deemed infeasible, so as to make it more presentable, the manufacturer was declaring to his CEO that Aramis was feasible, even though it was a bit complicated for his taste. While the company head is blaming himself—retrospectively—for having lacked courage and saying that he "ought to have" killed Aramis, on the other side the manufacturer is accusing himself—also retrospectively—of the same sin. He "should have" had the courage to stop the project. With perfect lucidity, the company head recognizes that he forced the manufacturer to complicate, reluctantly, a project he didn't believe in, while the manufacturer for his part admits that he agreed to complicate Aramis only because it was the shortest path to achieving it, given the strategic position occupied by the person who wanted to complicate it—a position that the manufacturer had given up trying to modify!

Two conversions, one definitive and the other tentative, transformed the Aramis project, which was at a standstill, into a political football. It was deliberately attached to the context of Right/Left alternation by the RATP company head. To this attachment, a Matra head had been reluctantly converted. As for Matra's president, he thought he had found a second VAL in an Aramis supported by the Left. This enthusiastic support was itself attributable to the RATP president and to the minister, both Communists, who thought they had, in an Aramis unanimously supported (they believed) by the technologists, a showcase that would present both high-level French technology and a renovated public-transportation system. How could they doubt Aramis, since both of the highest-placed directors of a private firm supported the project along with all the engineers of the RATP and the Transportation Ministry? Once again, as at Orly, Aramis was a dream, the ideal sociotechnological compromise, the dream that would simultaneously advance the P.C. (the Communist Party), the P.C. (the Petite Ceinture), capitalism, socialism, modernization, the maintenance of great social triumphs, and in particular would make it possible to bring about, in one fell swoop, both state-of-the-art research with multiple goals and industrial development that would transport real passengers.

"Things are getting clearer," my mentor declared. "Once again, you see, all is for the best. It could have worked. Aramis' road is paved with good intentions. We have to go check with the other partners, especially at the Transportation Ministry and the Ile-de-France Region, to find out whether they, too, believed in Aramis because the others did."

"But there's no end to this," I said to Norbert. "We'll have to go from the minister's office to the Minister himself, from there to the president, and from the president to all their international counterparts.

Why not go see Ronald Reagan, or the Chinese? And why not follow the chips clear to South Korea? After all—they, too, decide on Aramis' fate; they're its context."

"We have to stop somewhere."

"So it doesn't matter where? We just stop when we're tired?"

"One, when we run out of money for expenses; two, when the contextualizers themselves stop. If they tell us, 'The minister was interested in it for only a few minutes; he had other things on his mind; he put me in charge of the dossier,' we can retrace our steps, since Aramis has fallen out of the minister's purview. Beyond that limit, the analyses are no longer valid, since they're no longer specific. Then we'd study something else—ground transportation, for example, or Communist ministers, or technocracy."

"But I thought we had to take *everything* into account. I even read some philosopher, I think, or a sociologist, Edgar Morin, who said that every techno-bio-political problem was also a political-techno-biological problem . . . and that the politics of chips were also the chips of politics or something like that. You're letting go while everything is tied together."

"Few things are coming together, on the contrary; they're rare and fragile filaments, not big bubbles to be tied together by big arrows. Their extensions are unpredictable, it's true; their length as well. And they're very heterogeneous. Maybe we'll go to South Korea after all, or we'll go see Reagan, but simply because the Aramis maze will oblige us to draw a picture of that corridor of its labyrinth, and because an Ariadne has slipped her thread into it, not because we have to take into account the international element, or the technological infrastructure."

He even obliged me to observe for myself that the violent blow he struck with his fist on his desk had no visible influence on the chapter of Aristotle's *Metaphysics* that was filed under the letter *A* at the top of his bookshelf.

"You see: not everything comes together, not everything is connected."

After that interesting physics demonstration, he harangued me again about the notion of networks. They were all fanatical about networks in that shop.

[INTERVIEW EXCERPTS]

M. Gontran, researcher at the Institute for Transportation Research:

"For Fiterman, the issue was modernization of transportation systems. Aramis had a certain public appeal; it was good for export. These things made it possible both to develop state-of-the-art technologies and to make the ministry look good.

"Beyond that, Lagardère and the Communist minister intrigued each other.

"Obviously, it was more important as a project for Quin than for Fiterman. Fiterman invested very little in Aramis. In comparison, he spent a year and a half fighting for the A-320; in the end, Aramis was just one more thing. But for Quin, it was important. It was presented to him as something at the cutting edge that could be implemented quickly. They cut all the corners; but since it was an old project, they couldn't say that they'd still need four or five years more to develop it!" [no. 42]

A former member of Fiterman's cabinet, M. Marin, also a Communist, who no longer works in transportation, showed us the very dossier Fiterman had on his desk when he was making his decisions.

"You know, Fiterman had a kind of proletarian common sense. Even if his friend Claude Quin was singing Aramis' praises, Fiterman wanted to see for himself.

"I have one more note of his on the Aramis dossier" *[brandishing it]:*

[DOCUMENT]

Before deciding, sum up pros and cons on this matter after getting opinions from all "interested" parties. Seek out reliable and impartial opinions, *if possible!* —C.F.

"His own underlining! You see that he's not naive. The entire technology lobby was in favor. That's why I'm astounded that you're coming to see me to find out why Aramis died.

"I called several meetings to verify the degree of commitment of everybody concerned; we did several polls. *Everybody was for it.*

"The R&D effort was very attractive. I'd really like somebody to convince me that *it wasn't feasible,* with all those branches that gave it excellent coverage and that let it go get customers and bring them back onto the standard metro lines.

"It's true that there were security issues. A woman alone inside with a rapist . . . We had some concerns . . .

"But *if the RATP was ready* to put millions on the table, it's because they believed in it. And Matra, too, in the private sector. I didn't have any independent way of doing a technology assessment in my office; *if people coming at it from different logical angles came to identical conclusions,* what could *I* say?" [no. 43]

"It's like the Orly phase: everybody is unanimous about Aramis, but, as in a poker game, nobody thinks the others are bluffing."

"Still, that hardly makes it worth our while to do 'refined' sociology, Norbert. Say what you like, but the politicians have gotten themselves all worked up—it's as simple as that. They've forced the engineers to do things they didn't want to do. Even your famous principle of symmetry is no use; we've got exactly the opposite situation here—politicians distorting a technological logic that was perfectly clear. It's infeasible. I've felt that myself all along, but of course I'm an engineer."

"Of course, of course . . ."

During a given period, the form, scope, and power of the context change for every technological project.

Aramis' contextualizers tied in the major projects of the left with decisions about the variable-reluctance motor, nonmaterial couplings ("interesting for forking points"), and automation ("to prepare people's minds"). Very good; but this work refrained from adding to the mix a number of other players that other contextualizers, interested in other projects, are about to extract from the context and reassemble as allies or enemies of their business. At the same moment, with the same Left holding the same powers, the European airplane, the Ariane rocket, the Poma-2000, the Val-de-Marne tramway, the Rafale, TRACS, VAL, though very close, are in other contexts.

That is why the Context is such a bad predictor of a project's fate, and why the tedious argument over "individual freedom" and "the weight of structures" does not allow us to understand Aramis. The company head, the industrialist, the director of technological services—any of these people could have decided, or not decided, not to bring to bear on Aramis the weight of the Left, or of Technological Evolution, or of Necessary Modernization, or of the Equipment of the South Paris Region. They all could have decided not to let these forces of different origins get mixed up with Aramis' fate. Where is the freedom of the individual actors? Everywhere, in all the branchings of the context. Where is the structure? Everywhere, traced by all the branchings and relationships of the context.

Still more foolish, the quarrel between the history of the contingent bifurcations and the sociology of the structural necessities is of no use at all. To do contingent history is also to structure, to contextualize, and thus to gain or lose in necessity. For Cleopatra's nose to have any bearing upon the battle of Actium, you still have to have a Roman general and an Egyptian princess attached by bonds of love, and don't forget the serpent coiled up in fruit cups. All the attachments to context are so many exquisite corpses.

[INTERVIEW EXCERPTS]

M. Coquelet, a transportation official, is speaking in his office in the Ile-de-France Region:

"It's a project from the culture of the Sixties, transporting people in a private Espace instead of in public conveyances . . .

"Now, obviously, we're culturally out of phase. But in 1984 a Communist minister wanted to increase *pure research independently* of economic factors; he wanted to showcase his enterprise so as to polish up its image with cutting-edge research."

"But I don't understand. Isn't it really very applied, the CET? Isn't it seen as research?"

"Yes, but don't forget that at the time people thought they could apply it to the World's Fair; then they said it would serve as a transportation system. But after the project was abandoned, the costs turned out to be considerable.

"The contract of the State-Region Plan is an addition, a compromise if you like, between the enthusiasm of Fiterman and Quin *[on the Left]* and the enthusiasm of people from the Ile-de-France Region, such as Giraud and Fourcade *[on the Right]*.

"Aramis was supposed to be a matter of compromises, like the Val-de-Marne tramway, with less reticence about Aramis because of the research aspect. In the Plan contract, Aramis was provided for in the document that confirms the commitment, under the heading of transportation and traffic. It isn't on the list of "extensions of the metro"; it has its own rubric, but it isn't under research, either—we don't have that. It's not the Region's job to finance research.

"I think that *at the top they didn't see the skepticism of the technological levels*. I personally was very skeptical. My president was very skeptical.

"At Matra and the RATP, they had the impression that the engineers *were unanimously defending the project*. They told us that it was going to work, so you can understand that when they came to find us later *[in 1987]* and told us, 'We've used up the money,' when they hadn't even gotten to the testing phase, we said, 'No, we've got to cut our losses.'

"I never had to convince Giraud to give it up, since *we weren't enthusiastic* anyway. It was a compromise.

"The snag was a pretty big one, anyway. We'd spent 100 percent of our budget, and *we were in the middle of the river without any idea how deep the ford was*.

"In any case, the Region's commitment was a compromise; we had insisted on specific sites and we had accepted Aramis as a compromise, so *as soon as there was a problem, we didn't put up a fight*. The project wasn't supported by anything but an idea on the part of some technologists.

"Then the beast had to be *snuffed out*. It wasn't a 'filthy beast,' of course—but it was a Rolls Royce. I may have been a little harsh, but Aramis is a bit old hat, after all—a bit outdated. We need *mass* transportation, not individual transportation." [no. 34]

"You see?" Norbert remarked. "He tells us that it's a technologists' idea, and in the same breath that the people at the top didn't see the technologists' skepticism. Really, we need an even more refined sociology to explain this story . . . Did you hear what he said? The 'filthy beast' that has to be snuffed out? It's like the story of Frankenstein. In Mary Shelley's book, you don't know which is the monster that has to be crushed—whether it's the master or the frightful thing he's concocted and then left behind."

On an Alpine glacier, Victor encounters his creature, who seeks to explain why he has become wicked after being abandoned:

"How can I move thee? Will no entreaties cause thee to turn a favourable eye upon thy creature, who implores thy goodness and compassion? Believe me, Frankenstein: I was benevolent; my soul glowed with love and humanity: but am I not alone, miserably alone? You, my creator, abhor me; what hope can I gather from your fellow-creatures, who owe me nothing? they spurn and hate me. The desert mountains and dreary glaciers are my refuge . . ."

"Why do you call to my remembrance," I rejoined, "circumstances, of which I shudder to reflect, that I have been the miserable origin and author? Cursed be the day, abhorred devil, in which you first saw light! Cursed (although I curse myself) be the hands that formed you! You have made me wretched beyond expression. You have left me no power to consider whether I am just to you or not. Begone! relieve me from the sight of your detested form."

"Thus I relieve thee, my creator," he said, and placed his hated hands before my eyes, which I flung from me with violence; "thus I take from thee a sight which you abhor. Still thou canst listen to me, and grant me thy compassion. By the virtues that I once possessed, I demand this from you. Hear my tale; it is long and strange . . ." [From Mary Shelley's Frankenstein*]*

Why reject me? Have I not been good? Was I not born well-endowed with virtues, unlike my brother VAL? Have I not been the dream, the ideal? What pains were not taken for my conception! Why recoil in horror today? Did not all the fairies hover over my cradle? Oh, my progenitors, why do you turn your heads away, why do you confess today that you did not love me, that you did not want me, that you had no intention of creating me? And if you have given me existence, why do you take it back again so soon? And if you did not want me, why did you keep me alive, year after year, in that glacial limbo, attaching to me dozens of poor devils who sacrificed their nights and their ardor to me? If I have been badly conceived, why not conceive me again? Why not take the trouble to reshape me? Why do you turn your heads away? Am I a Medusa, then, I whom you so loved? Who has committed the inexpiable crime of abandoning a creature drawn out of the void? I, who did not ask either to be born or to die? Or you, who insisted that I be born? Of all the sins, unconsummated love is the most inexpiable. Burdened with my prostheses, hated, abandoned, innocent, accused, a filthy beast, a thing full of men, men full of things, I lie before you. Eloï, eloï, Lama, lama sabachthani.

1. RATP advertisement: "Darwin was right"—meaning that buses transform themselves like biological species by adapting more and more closely to their environment. But this evolutionary mythology leaves human society out of the picture and does not provide any way of accounting for the sometimes cruel fate to which technologies are subjected. [Photo Bruno Latour]

2. The Aramis car in the former foyer of the Matra Transport company in the suburbs. Aerodynamic and elegant, the car is smooth and shiny; but it's inert. A few months after this photo was taken, it was sent off to the scrap yard. Here is an isolated technological object turned into a museum piece—as incomprehensible as the buses in the photo above. [Photo RATP]

3. In 1973, in a beet field near the Orly Airport runways, the first cars of the first Aramis prototype were tested. This photo shows the test track, along with two little cars that are running quite close to each other, even though they are not mechanically coupled. [Photo RATP]

4. May 3, 1973: Aramis is unveiled to journalists. The four-seat cars do not resemble the elegant one in Photo 2, yet the basic principles of the Aramis system will undergo very little change over the next fifteen years. [Photo RATP]

5. The Aramis carcass as it could be seen during the Orly phase. The arms of the on-board steering mechanism allow each cabin to branch off individually; the electric motor will later be replaced by the rotary engine. [Photo RATP]

6. Still at Orly, in December 1980: the Aramis car now has eight seats, a nice yellow color, and its famous rotary engine. The relaxed appearance of the passengers is misleading: this vehicle is still a prototype that continues to run in circles in the same beet field shown in Photo 3. [Photo Matra]

7. The control room for VAL in Lille. The success of big brother VAL cast a shadow on little Aramis. A major innovation: the former drivers are now positioned in front of a control panel, and regulate the movements of the automated trains from a distance. [Photo Matra]

8. Aramis is finally going to exist: in July 1984 the great men of its world stand around a model congratulating one another before signing the agreement to go ahead with full-scale experiments. From left to right: Claude Quin, RATP president; Michel Giraud, president of the Ile-de-France Region; Jean-Pierre Fourcade, vice-president of the Region; Charles Fiterman, Minister of Transportation—who will resign a few days later; and Jean-Luc Lagardère, CEO of Matra. [Photo RATP]

9. On the boulevard Victor along the Seine, next to the quai de Javel, which is being reconstructed, the Center for Technological Experimentation (CET) is rising from the ground in early 1985; later it will connect with the abandoned Petite Ceinture. The Eiffel Tower stands out above (center). The workshop is at the very bottom, and the experimental station (see the map in the frontmatter) is at the far left. [Photo RATP]

10. The construction of the experimental site is well under way. This photo was taken from the eastern side of the workshop. Beside the tracks the guide rails and electric power supply rails are clearly visible, along with the guides for the rubber tires. [Photo RATP]

11. Aramis really does exist now; here it is in 1987, pulling up to the experimental station on the boulevard Victor. Each pair is made up of two ten-passenger cars, coupled mechanically like standard train cars; but the only connection between pairs is the electronically calculated "nonmaterial coupling." [Photo Matra]

12. The nonmaterial coupling allows passengers who are heading for two different destinations to get on at the same station. When they leave the station, the two pairs of cars proceed together, adjusting the distance between them through the principle of the CMD (adjustable mobile sector) like cars in a lane. [Photo RATP]

13. After the two pairs of cars have passed a branching point, each pair goes its own way, thereby feeding—without transfers—the transportation networks that a traditional line would be unable to serve. [Photo RATP]

14. Aramis' Central Command Post. As with VAL (whose control room is shown in Photo 7), the wholly automated system is not intended to have any human actors except at the two extremes: the passengers riding in the cars, and the controller at his desk. [Photo Matra]

15. In March 1987, seven months before its demise, Aramis is still inspiring enthusiasm in the ministers of the cohabitation period, Michèle Barzach and Edouard Balladur, along with the prime minister himself, Jacques Chirac (at right), who jokes about the nonmaterial coupling between cars: "I've never tried it myself." [Photo RATP]

16. One of the cars of pair number 4, suspended between heaven and earth in the workshop on the boulevard Victor, at the precise moment when the development program stops. It is in this already terminal state that the investigators meet Aramis for the first time. The inquest into Aramis' death is about to begin. [Photo Matra]

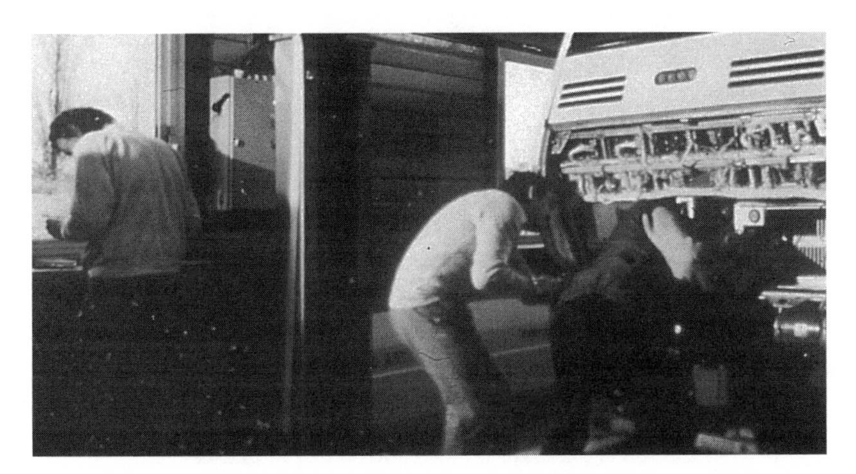

17. Aramis in its frantic last days. Technicians surround its carcass, trying to debug its program and make it work automatically—that is, without their constant intervention. Depending on their success, the car will be swarming inside with bits of information or outside with technicians. [Photo RATP]

18. Aramis' revolutionary engine, which was the key to the realization of the project. The idea was to take the axle of the wheel as the rotor of the electric engine. This made it possible to dispense with gears and transmissions and allowed for very fine calibration of the wheel movement; this, in turn, made it possible to calculate distance accurately. [Photo RATP]

19. Since no engineer could prove mathematically that the cars would not hit each other, telescoping bumpers were added—a perfect case of a compromise among mathematical modeling, engineering knowhow, and the traditions of automotive and public transportation. The system became increasingly complex as it was forced to absorb more contradictory constraints. [Photo RATP]

20. The sad situation of the building in which Aramis had originally looked forward to a bright future. This is the same space as the one shown in Photo 16. Now derelict and abandoned, it has become the province of tramps and graffiti artists. [Photo Stéphane Lagoutte]

THE 1984 DECISION: ARAMIS EXISTS FOR REAL

After the interviews, which were always exhausting because they demanded our full attention, we often collapsed in the nearest café.

"So we know that everything happened during that three-and-a-half-year interphase." My mentor was recapitulating on the paper tablecloth. "We also know that the mystery is not the death of Aramis but its rebirth in 1981, since we've eliminated the influence of context. Then there's this mysterious three-and-a-half-year delay, which we know was because of the upcoming World's Fair, and then because the World's Fair was dropped. Here again, Aramis should have died. Okay?"

"But Aramis goes on as if nothing had happened. It's intact, or almost."

"Yes, that's the only mystery; the CET isn't signed until July 1987. To account for this survival, this delay, we have two elements: up above, in the higher spheres, everyone is now in favor of Aramis, unanimously. Although everybody has private doubts about the project, they give it their own backing, however half-heartedly, because they see all the others supporting it enthusiastically. Down below, with the technicians, everybody is skeptical. . ."

"At least that's what they're saying now. At the time, no one noticed the skepticism . . ."

"Exactly. Everybody was skeptical, but only in private. That's the whole problem: half-hearted enthusiasms come together on high, while down below half-doubts are all scattered, isolated, buried in notes that we are often the first to see, in any case the first to bring together as a whole. What's more, every time someone up above asks somebody down below for an opinion . . ."

"The person up above gets an opinion that's more positive than the one the person down below really holds, because the people down below revise their opinions so they'll jibe with what they think the people up above really want . . ."

"Exactly."

"Because, between the top and the bottom," I added, making my own head reel with sociological analyses, "there must be people, intermediaries, who do the translation, who transform the technologists' doubts into near-certainties. So the decisionmakers think the thing is technologically feasible in addition to being politically opportune; and the same intermediaries transform the decisionmakers' fears into near-certainties, into orders given to the technologists. So the technologists think Aramis has political support."

"Ah! So it's the technocrats, is it? They'd be good villains . . ."

"Or else it's just what I was trying to tell you, Norbert: they all got carried away. If I were a journalist, I'd go tell the whole story to the *Canard enchaîné**. They kept on going with a project that defies common sense."

"No, you're wrong about that—it's a perfectly normal project. The *Canard* would have nothing to sink its teeth into. And that's just what's bothering me. It's a muddle, but it's not unusual enough to explain Aramis' survival or, later on, its death. The technocrats are in the same place, doing the same job of translation-betrayal in all the successful cases: VAL, Ariane, the Airbus, the Poma-2000, the tramway. No, they could have brought it off. In any case," Norbert added threateningly, "you're not to say a word to a living soul about this without my permission; you've signed on the dotted line."

"I was just kidding . . . Still, the fact remains that there are only two solutions. Either they're all incompetent, or else there's someone who has a clear strategy and who's pocketing the cash. If you rule out the first solution, you have to look into the second. It didn't happen by

*A satirical weekly that specializes in uncovering secret scandals.

accident, after all. There has to be somebody in this story who's making a profit. Or else I've had it right all along: they've all gotten carried away. In either case, if I were a journalist—if I *were,* but of course I'm not—there would be something for the *Canard.*"

"Strategy, my good man, is like context: an invention of vulgar sociologists."

"Do you mean to tell me we're going to do 'refined' sociology again, just when good old-fashioned crude sociology would be all we'd need to denounce a good old-fashioned crude scandal?"

"Not refined, *hyperrefined* . . ."

And we went back to our interviews to "explain" the famous decision of July 1984, giving up the helpful, handy solution of denouncing the technocrats.

"The only actor that may have had a strategy was Matra. Remember what Girard told us."

[INTERVIEW EXCERPTS]

M. Girard:

"In 1983, Matra was *pushed.* They couldn't go against the left-wing administration, but at the same time they kept coming back to VAL.

"In fact, Aramis was *imposed* on Matra *against the will* of its leadership team, which doesn't mean that they didn't do a good job; on the contrary, it worked—the results are in—but *they didn't believe in it.* They *said to themselves:* 'For 30 million, *if that's what they want,* okay, we'll do it, and it'll bring in 150 million in research money.' But *VAL was always their target.*

"They remained wedded to the old standbys. After Bardet, there was nobody who was prepared to pick up the somewhat radical innovations and run with them. Matra *never went back to Aramis,* in the end. They *stuck with* the classic standards, judiciously applied." [no. 18]

M. Frèque, who headed the Aramis project at Matra after the new startup is speaking with the same frankness, the same subtlety in his analysis, the same extreme cordiality. The Aramis cab is still on display in the entry hall.

"Let's say that we had the will to do it, but . . ."

"But not the way!" *[Laughter]*

"That's exactly right: we didn't have the way!" . . .

"The RATP *said to itself,* 'I'm behind, Matra's ahead of me, I have to do something better than VAL.' We proposed a simplified Aramis, an Araval; it was in a note in 1982, an internal memo. I still have it. The RATP reacted very badly." *[He reads excerpts from the internal memo, which he doesn't want to hand over.]*

[DOCUMENT]

M. Maire got a very bad impression. We backed away from developing Aramis . . . If we do a VAL derivative, they won't do anything . . . Have to *change our language* at Matra . . . *Not give* M. Quin *the impression that we want* to downgrade the system. What *motivates* M. Quin is a *sophisticated system.*

"At the same time, you can't say that simplification would have solved everything, since the cost is in the infrastructures in any case. But when you get right down to it, what you have to see is that the RATP *wanted more* than VAL.

"On our side, *we didn't see the technological imperative,* but it was clear that *the client wanted it to be complicated.* Again, doing away with train configurations didn't solve the problem. There's no point exaggerating, after all. There really wasn't any method, but we had to find a compromise.

"We had to find a structure and an organization that would allow a compromise.

"The operating agency overspecifies; there's a normal, ordinary overkill built into technologies and specifications. And on the other side of the fence, the industrialist tries to do as little as possible that isn't specified. He'll say, 'It isn't written down; I'll take that off the shelf and that's the end of it.'

We had to find a compromise. On VAL, we were 80 percent there *[see M. Frèque's description of the VAL negotiations near the end of Chapter 3].* But for SACEM or for Aramis, there were a lot more problems." [no. 6]

The actors don't have a strategy; they get their battle plans, contradictory ones, from other actors.

The actors in a technological project populate the world with other actors whom they endow with qualities, to whom they give a past, to whom they attribute motivations, visions, goals, targets, and desires, and whose margin of maneuver they define. It is precisely because of this work of populating that they are called *actors.* For a given actor, this is the way the *strategy* of the other actors is interdefined. What does Matra want? It's M. Girard, head of the RATP, who says what Matra wants. According to him, because of its past, as builder of VAL, Matra is clinging to its own way of being and no longer wants Aramis, in which it has very little faith; now it wants VAL. What does the RATP want? According to the head of the Aramis project at Matra, the RATP finds itself lagging behind Matra and wants to catch up, presumably to get even. From this attribution of a past and a feeling, Frèque deduces a behavior by applying a rule of continuity between the past and the future: the RATP doesn't want a project that would resemble VAL; only a sophisticated system is worthy of its interest. How much maneuvering room does Matra have, as the RATP sees it? Not much. From Matra's customary behavior—the behavior of a company capable of producing VAL—is deduced a tendency, a weight, an attractor: make Aramis a mini-VAL. Why not make it an Araval? Because Matra doesn't dare displease the left-wing administration, and its director has bound up Aramis' fate with the Left. How much maneuvering room does the RATP have, in Matra's view? Not much. The RATP has the desire to produce something other than a mini-VAL, at any price. The attractor, here, is VAL as foil. Does there really exist a causal mechanism known only to the sociologist that would give the history of a technological project the necessity that seems so cruelly lacking? No, the actors offer each other a version of their own necessities, and from this they deduce the strategies they ascribe to each other.

What are the strategies that the two actors can deduce on the basis of their own reconstructions of the motivations and maneuvering room of the other actors? Matra draws the conclusion that "it's especially critical not to give M. Quin the impression that we're pursuing an Araval under the guise of an Aramis"—that Matra has to "change its language," has to dissimulate its "real goals," which are to take on the fewest possible complications while aiming at Araval. The head of the RATP concludes that, once desires have been inscribed in Matra's nature, nothing more can be done. Aramis has been in a state of abandonment since Bardet's day. Among the possible results of this interdefinition of the past, of motives, goals, and more or less indirect means, it turns out that the two interviewees agree. "They're aiming at VAL in spite of the complicated Aramis that we wanted to impose on them." "We were aiming at VAL in spite of the useless complications that they wanted to impose on us." This superposition is an exception. The methodological rule that consists in letting the actors define each other can accommodate all cases, including the miraculous one of agreement between the one who is defining and the one who is being defined.

"But all these interviews, Norbert—we're doing them in 1987 and talking about 1981, 1982, 1984. How can we believe these people? And besides, the fact that the project has been terminated makes them even less believable. They all know what happened. They can all tell us stories about ineluctable destiny; it doesn't cost them a thing."

"They're all telling us stories, that's for sure. But they all tell themselves stories, stories about strategy, scenarios, with lots of things like: 'Once upon a time there was an RATP that wanted to get even for being humiliated'; or 'Once upon a time there was a charming capitalist who dreamed of entering Paris right under the king's nose.' We have to write everything down, that's all. Who's saying it? About whom? To whom? When? Referring to what period?

"So you write it down:

"M. Frèque, on such-and-such a date, tells us (that's you and me), speaking about 1984, that in his opinion the RATP, represented by M. Girard, wanted such-and-such a thing and that he learned it by undergoing such-and-such a trial.

"Look at the note he didn't want to let us have. That's an experiment. He thought the RATP was flexible and ready for a compromise. He notices that he was wrong. That the RATP was furious, that Matra had to change its language. He tells us, *us,* that this experiment revealed to him the true goals and inclinations of the RATP."

"But is it true? Did it really happen that way? Did the RATP really want that?"

"We don't know a thing about it, and that's not the issue. All we do is write down the stories people tell us."

"But you're not being asked to write a novel; you're supposed to provide the truth. That's what they're paying you for, Norbert."

"No. First of all, I'm the one who's paying you to help me, and I'm paying you to write everything down—all the stories of goals and desire and trials. Not so you'll unearth the truth in the actors' stead. The truth will come out of the novel, out of all the novels told by all the interviewees about all the others."

"A total novel?"

"Not even; we'll let the actors do the totals themselves."

"Several different totals? When I think that I could have spent the year doing real technology, that I had a good slot in the Man-Machine Interaction and Artificial Intelligence program!"

"But you're there, my friend—you're up to your ears in it. This is what man-machine interaction is, and artificial intelligence."

"Sure it is! Concocting a novel about people who're writing technological-fiction novels and leading one another down the garden path!"

[INTERVIEW EXCERPTS]

On the boulevard Victor, in the project's now-abandoned offices, M. Parlat is expressing rather heated feelings about the manufacturer:

"Let's say that Matra took us for a ride. They never intended to go through with Aramis. When they saw the problems piling up, they chucked it.

"It was a case of a manufacturer pursuing its own interest; that's normal.

"We're rivals. Before this, the RATP always dealt with subcontractors.* But this time they were acting as general contractor, so the less we knew about it the better.

"It's private versus public. It's not a total loss for them, by the way, since they paid for all their research for Lyon and for SACEM out of the Aramis budget, the public budgets." [no. 2]

*The question of who is in control of technological competency is a crucial issue in all projects: the contracting authority is further removed from the details than the contractor. As for subcontractors, they simply arrange for the various specialized services.

"He agrees with me," I remarked modestly.

"Well, you ask Matra the next question."

[INTERVIEW EXCERPTS]

At Matra headquarters, still in M. Frèque's office:

"I'm sure you're aware that many people have floated a Machiavellian hypothesis about your strategy, suggesting that you financed your research with the Aramis contracts but that you were actually looking at something else, a second-generation VAL. What do you think about that?"

"There's *no such thing* as Machiavellianism. To be Machiavellian, you have to be very smart and work very hard for a very long time and be very single-minded. Well, there just aren't many people out there who are very smart and very hard-working and very stubborn . . . *[Laughter]*

"No, if you really look at it, from Matra's point of view Aramis is a financial catastrophe. We went over budget; we didn't earn anything at all.

"What happened was that, on the contrary, we *really believed* in Aramis. Turning the CET into a research center was a step backward.

"What happened was that we had doubts about VAL. We paid too much attention to the other side, to people telling us: 'It'll never sell'; 'It was good for Lille but not for Bordeaux or Brescia.'

"And VAL was too expensive from the point of view of infrastructure for a lot of medium-size cities. So, no, on the contrary, Aramis was cheaper, could be cheaper. We really needed Aramis; it *completed our product line*—that's why we were so determined to end up with a line.

"We had doubts about VAL. We were too pessimistic after being too optimistic, so much so that in Strasbourg we initially proposed Aramis instead of VAL . . .

"In the end, in spite of the costs, it turned out that Strasbourg *wanted* VAL.* It wasn't the price tag that mattered to them; it was the relation between price and public image. For them, VAL wasn't a gimmick; Aramis was, in spite of everything." [no. 41]

*"In the end" is as reversible in technology as in politics. The 1989 municipal elections were fatal to VAL as well as to Aramis. In Strasbourg it was the tramway that won out.

The actors create both their society and their sociology, their language and their metalanguage.

Not only do a project's actors populate the world with other actors, but they also define how the populating will come about and how to account for it. M. Girard, of the RATP, has his own ideas about social physics: "Matra," he says, "was pushed"; it floundered; it didn't act; it was a patient, not an actor. He also has ideas about what is possible in France—the manufacturer "couldn't" go against the left-wing administration.

The RATP engineer, M. Parlat, has quite explicit views about the *interests* that push or pull a manufacturer to profit from the weaknesses of the powers that be.

The father of VAL develops very precise sociological theories: Machiavellianism is impossible for want of Machiavellis who are hard-working enough and stubborn enough to stick with a strategy long enough. In contrast, manufacturers are at the mercy of the "what-will-people-think?" phenomenon, and they internalize others' doubts about their own capacity to accomplish major projects; they shift too quickly from optimism to pessimism; they, too, are fragile and cyclothymic. As for provincial cities, they cannot choose a solution by looking at profitability alone, since their principle for choosing is a relation between "price and image" offered by a public-transportation system.

There are as many theories of action as there are actors.

Does there exist, after all, a theory in which all these actors and all their theories can be summed up, one that would enable the sociologist-king to speak with some authority?

That depends again on how the actors act to disseminate their own theories of action. Can Girard's social physics be extended to the point of interpreting the others? Is it, on the contrary, the doctrine of the private-enterprise-that-pursues-its-own-interest-at-the-expense-of-the-public that will prevail and encompass the others, which are then accused of Machiavellianism? Or will the winner be the doctrine according to which all Machiavellianism is impossible, owing to the weakness of human nature and the uncertainties of economic calculations—in which case those who level charges of Machiavellianism are themselves charged in turn with evil intentions?

To the multiplicity of actors a new multiplicity is now added: that of the efforts made to unify, to simplify, to make coherent the multiplicity of viewpoints, goals, and desires, so as to impose a single theory of action. In the strange arithmetic of projects, everything is added; nothing is taken away, not even the rules of the metalanguage, not even the arithmetic's variable rules by which addition and subtraction are defined!

"But it's in his interest," I said indignantly, "in Frèque's, that is, to tell us that. If he chooses a sociology of interest, he can't avoid being accused of Machiavellianism. So he talks to us about uncertainty, tinkering, pessimism . . . Another minute and he'd have had us weeping over the manufacturer's misfortunes; it was clear from the start. That's Machiavellian, isn't it? To say, out of self-interest, with your hand over your heart, that you don't have a strategy and that you're just a poor slob with nothing but the shirt on his back who's doing the best he can."

"And your interpretation, my dear sociologist, is based on what sociology?"

"Well, people tell us what it's in their interest to tell us."

"Of course; but do they know what's in their interest?"

"I'm not sure. But spontaneously, unconsciously, yes, they probably do. He's such a good negotiator, that one, he looks so clever, Frèque does, that I'd be astonished if he weren't quite up front in pursuing his goals."

"So you'd be ready to put your hand in the fire and swear that Matra really did take the RATP for a ride to get its own research funded, and that it never intended to go through with Aramis?"

"Let's say my little finger. I don't know enough sociology yet to put my whole hand in . . ."

"Your modesty is not to your credit. It's the height of arrogance."

"Arrogance?"

"Of course. You're an old positivist, in spite of your youth, and modest positivists are the worst of all. We'll never know enough sociology to judge the actors—never. They're the ones who teach us our sociology."

"And there are as many different sociologies as there are actors?"

"Exactly."

"As I was saying," I sighed. "I should have stayed in the Man-Machine Interaction program and AI."

"'Hee haw!' you mean, 'hee haw!' You're just like Buridan's ass, you are—you never know what you want . . ."

[INTERVIEW EXCERPTS]

M. Cohen, head of the Aramis project at Matra in the Orly days:

"They're 'conversational,' you know, the relations between people and technological projects.

"The personalities of the project heads have a lot to do with it. If Frèque had headed up Aramis at the start, the project would have developed differently, there's no doubt about it. And if I'd been the head of VAL, *it would have been a different VAL.*

There's no question that if you compare the Aramis project team and the VAL team (there were about ten people on each), you see that every project has a personality. One project reasoned in terms of redundancy and looser reliability: that was Aramis, automatically more creative. And on the other side, with more intrinsic security but with its creativity more stifled, more strict as well, more rigorous, there was VAL.

"The change of team accounts for a lot. The VAL team that took Aramis in hand *[after 1984]* overturned everything, and what they produced looked like VAL.

"What's more, we know something now that we didn't know at the time, namely how much a transportation system costs. *No matter how big Matra was,* they could not have done both at the same time. And they made a judicious choice: they came down in favor of VAL. That was realistic, since there were two concurrent systems."

"All that's left is the size difference between Aramis and VAL."

"Yes, the size difference remains." [no. 45]

To study technological projects you have to move from a classical sociology—which has fixed frames of reference—to a relativistic sociology—which has fluctuating referents.

If the actors in a project define not only the other actors' essences and desires but also the rules for interpretation that make these definitions workable, all their viewpoints have to be deployed in a supple frame of reference—*a reference mollusk,* as Einstein called it in his essay on relativity. Since they themselves establish their theories, their metatheories, and even their metametatheories, the actors have to be left to their own devices. It's a *laissez-faire* sociology.

Not only does Cohen, the former head of the Aramis project, say that team culture and the preferences of project heads influence technological choices; not only does he reconstruct possible stories (the story of a VAL that he would have directed in Frèque's place, the story of an Aramis that he would have pulled off); not only does he blend cultural and psychosociological determinism (creativity on one side, rigor on the other) with technological choices (probabilistic security on one side, intrinsic security on the other); but in addition, during the interview he changes his rule for interpretation, moving from people to organizational and economic necessities ("in any event, Matra could not have done both at the same time"). And all this in a interview fragment that lasts two minutes and twenty-five seconds! What's more, the observer must now compare these variations with the other interviews—for example, the one with Cohen's old classmate Frèque. "We really wanted to make an Aramis that would be different from VAL," said Frèque. Cohen's answer: "Even if they had wanted to, they couldn't have"—owing to organizational necessity (they're too small), financial necessity (it costs too much), psycho-techno-cultural necessity (Frèque, influenced by VAL, will never do anything but mini-VALs).

Is it impossible, then, to tell the story of Aramis if all the rules diverge, if the laws of sociology vary with the point of view and from minute to minute for a given point of view? No, because the actors also provide themselves with the means to pass from one point of view to another, and they unify, from their own point of view, and each for himself, the multiplicity of points of view thus deployed. Each constructs his own instrument in order to elaborate a synoptic view. All the actors thus repair, for themselves, the disorder they create by multiplying perspectives.

[INTERVIEW EXCERPTS]

M. Gontran, a technical adviser in Fiterman's cabinet, is speaking in his office at INRETS, with the same affability, the same sensitivity to subtle variations in technology and politics, the same modesty found throughout the whole world of guided transportation:

"Was Matra trying to get funds for research on other projects, or was it seriously interested in Aramis? Could its approach be called somewhat Machiavellian?"

"I don't know." *[Long silence.]* "I'm not sure if I can say anything very specific about it.

"What I do know is that Matra *always* contemplated going beyond the R&D stage to the production stage. After Aramis, they considered selling a mini-VAL.

"No, I think they were *really* concerned about the manufacturing logic. However, the minister and the RATP *didn't have* the same production concerns; *there was a game of cat-and-mouse going on.*

"Matra didn't want to get in too deep, without a decision to begin production, and they were also *pushing* the dossier in order to get the line, for the sake of having a line. With the [the World's] Fair, everybody was pushing, but then the supports gave way and the project was left hanging . . ."

"That takes care of your interpretation," said Norbert on his way out. "The Machiavellian hypothesis doesn't hold water. They really tried to go through with Aramis, at least at first, and it was the RATP, on the contrary, and the public powers who were not so interested. Hence the game of cat-and-mouse. So Matra is the mouse that gets eaten."

"No, no, it's the cat. Matra is doing the manipulating; it's playing with the RATP."

"Will your interpretation stand up to the test of the interviews or not? That's the only criterion. Everything hangs on the business of the line. If Matra is pushing Aramis, the CET, and the Petite Ceinture line simultaneously, it means they really do want to do Aramis. If you can prove to me that they only want the CET to pay for their own research but don't want the line, then I'll start believing in your Machiavellian hypothesis."

With a technological project, *interpretations* of the project cannot be separated from the project itself, unless the project has become an object.

The actors' or observers' interpretations of the actors' motivations and interests become more real and less real as a function of the progressive realization or derealization of those interpretations. Frèque attributes *intentions* to his CEO, to the company head, to the RATP, to nonmaterial couplings, and to variable-reluctance motors, just as he attributes rules of behavior to provincial cities, to France, to the private sector, to the public sector, and to humanity in general. He lines up the actors, humans and nonhumans alike, in a narrative; he mobilizes them in a scenario in the course of which Aramis exists for real on the Petite Ceinture; he offers them roles, feelings, and ways of playing. He creates a whole world, a whole movie, a whole opera. Will they follow along? Will they play with him? If the actors lend themselves in large numbers to what Frèque expects of them, then his interpretation of their roles as well as the Aramis object that they're charged with creating will both be realized.

But some of them protest. "Intentions are being attributed to me!" they cry indignantly. "I never wanted to pursue Aramis beyond the World's Fair project," say the public authorities. "It was never a question of anything but research on automation," say the researchers. "You're giving me the role of the Rat, but I want to play the Mouse." "You don't know those people—they're Machiavellian." And there goes Aramis, moving *backward* along with the interpretations of one of the screenwriters. The actors, disbanded, are recruited by other screenwriters, given new roles, dressed in new costumes, entrusted with new scripts, and there they go again . . .

So translation is not the starting point for an action, but the first result of a preliminary scenarization. To make someone *deviate* from her goals, that someone first has to be defined; goals have to be attributed to her, a social physics has to be proposed that makes her susceptible (or not) to deviation, and a psychology has to be proposed that will make it possible to explain the deep feelings of the being in question, whose desires are then translated. Without that preliminary work, translation would be impossible. There would be actors on the marked borders who would know what they wanted and who could calculate the path that leads to their goals! There would be well-defined social groups endowed with well-understood interests! The world would be rational and full—and technologies would therefore be impossible!

[INTERVIEW EXCERPTS]

M. Gontran again:

"It's paradoxical, though, isn't it, that the project didn't budge after the World's Fair fell through?"

"You know, there's such a thing as an *announcement effect:* when you've spent a year and a half selling a project to the press while you were actually just shifting from research to development, you say, 'It's going to get done,' 'It's getting done.'

"Nobody *took responsibility for shutting down.* The CET was a way to stay in a holding pattern. Also, there was heavy *pressure* on the technicians: the merger technology, the nonmaterial couplings, the dense network, the variable-reluctance motor—all that was very *seductive.*

"Besides, at a time when a lot of innovations were being snuffed out, this was one of the few innovative projects in the field of transportation that the public could *see.*

"None of the other stuff *sells* politically; you can't have a ribbon-cutting ceremony to inaugurate it. All the political types had their pictures taken in Aramis" *[see Photos 8 and 15].*

Technological projects are deployed in a variable-ontology world; that's the result of the interdefinition of the actors.

In a thirty-second portion of an interview, a single interlocutor offers several theories of action. The same project which was "pushed" by the actors becomes impossible to shut down. So here we have a physical model. Aramis is a rock—of Sisyphean proportions—that has made it to the top of a slope owing to the work of human beings, but that then rolls back down the slope; human beings are powerless to stop it. It's a ballistic missile like the ship's cannon Victor Hugo described in *Quatre-vingt-treize:* a cannon that can't fail to roll right over those who might have the foolhardy courage to oppose it. But is the "announcement effect" also a physical model? No. The same Aramis project is awaited by everyone who has read in the newspapers that it was going to get done. Now we're in the social, or psychosocial, realm. But also in the juridical realm, since all journalistic publicity prepares people's thinking and forms habits that cannot be undone without talk of "false advertising." Already, lines of passengers impatient to show up at Bercy are forming on the boulevard Victor; "Aramis," they whisper, "is almost ready." So it's difficult to go back on one's word and announce, "Move along! There's nothing to see here except research projects." For the announcers' responsibility is on the line; they can no longer retreat. With the "pressure" of the technologists who add their power to the missile that's rolling down the hill, have we returned to the physical model? No, because their pressure originates in Aramis' "seductiveness." Technologists are enamored of the "merger technology" and even more so of the "nonmaterial couplings" that attach them to their project. So here's a real monster. A force that is both physical and amorous—a true Minotaur. But the interviewee doesn't stop there. He moves on to anthropology. How could someone who has already "snuffed out" so many innovative projects have the nerve to assassinate the latest one, tender and charming as it is? Wouldn't he be immobilized by *shame?* All the more so because Aramis can be "seen" by the public; it can be inaugurated. What model is implied here? A good model of political vanity: we love those who solemnly inaugurate what they have done with our money, in our name, and for our benefit. But this model of visibility is in turn interpreted thanks to the application of a commercial interpretation to the political world: "An innovation that the public can't see won't *sell.*"

Let's calculate the *sum of forces*—using this expression to designate both the work all the actors do to sum up and the diversity of the ontological models they use. Let's add the thrusts of human labor, the fall of ballistic missiles, the responsibility of promises, amorous seduction, the shame of more killing, vanity, business—everything that makes Aramis impossible to suspend. Yes, it's definitely a strange monster, a strange physics. It's the Minotaur, plus the labyrinth, plus Ariadne and her thread, plus Daedalus, who is condemned to die in it and who dreams of escaping. They're really fun, those people who write books in which they think they're castigating technology with adjectives like smooth, cold, profitable, efficient, inhuman, irreversible, autonomous! These insults are qualities with which the engineers would be delighted indeed to endow their hybrid beings. They rarely succeed in doing so.

[INTERVIEW EXCERPTS]

M. Gontran again, after a long reflective pause:

"Matra's strategy? I think it was pretty simple.

"Matra pulled off a tremendous coup with VAL, which really started to get off the ground in 1983, but they had some trouble selling it outside Lille; you mustn't forget that it was implemented in the United States in 1984, 1985, 1986.*

"So Matra went through a *dry spell* that obliged them, out of *concern* for survival, to look for some diversification, and Aramis was a way to *make sure Matra Transport would keep going*. Without being threatened exactly, Matra's future wasn't a sure thing, at a time when Lagardère was having to restructure his activities . . .

"Matra *really needed* to relaunch Aramis in order to diversify, to *maintain* its image of being in the technological forefront, and to *get hold of* public funding.

"Then I think there was a *change of strategy* at Matra, when they succeeded in Toulouse in 1985, I think it was; after that, the structure of Matra Transport was *stabilized* and their R&D component became relatively *less important*.

"The *headlong rush* into technology that allowed them to survive on whatever research funding they could get was no longer *necessary to their survival*.

"They touched all the bases of a policy that was fairly interventionist at the time, and they were *aware* of their superiority. And it's true that in terms of technological competence, they had an advantage over traditional railroads.

"Matra, don't forget, was the only company in the field that was hiring. It went from a small nucleus in R&D and technology to a *real* enterprise that had more normal and more traditional industrial *characteristics*." [no. 42]

To survive in a variable-ontology world, the promoters of a technological project have to imagine little bridges that let them temporarily ensure their stability.

Actors never swim twice in the same river. As they are defining one another, as they are changing ontologies and offering each other their theories of action, there's no guarantee of their own continuity in time. To say that the character of Hamlet in Act I is the same character we find in Act V, or that the RATP of Chapter 1 is "the same" as the RATP of Chapter 5, or that the Matra company is the same in 1985 as in 1982, you have to make an effort, impose interpretations, ensure continuity, recruit faithful

*In fact, VAL is being used only at Chicago's O'Hare Airport and in Jacksonville.

allies. Thus, not only is the actors' size variable, not only are their goals renegotiable, but their very *isotopy* is the product of work. For actors there's no such thing as the force of inertia, any more than there is for projects themselves.

For M. Gontran, Matra was really transformed after 1985. The company that loved Aramis was only a fragile being, the victim of a dry spell, that was involved more in research than in commercialization and that could survive only by a "headlong rush" into the frantic quest for sources of public funding. It was this particular being that "needed" Aramis so it could diversify, because Aramis was complicated and could therefore attract money from the government, which was always ready to support research. But the company, which finally succeeded in selling VAL outside Lille, now has less need for public assistance, and thus no longer needs to support complex projects; consequently, it falls back on VAL. Matra no longer loves Aramis, but it's no longer the same Matra. Not only has the company itself changed, but its way of *changing itself has changed.* It had been "rushing headlong" from project to project; now it's "stabilized," with "normal and traditional characteristics." Corporate bodies are no less fragile than the characters in a novel. To guarantee them some continuity from one end of a story to the other, you have to work like a dog, but the scope of this work may diminish in the course of the narrative. The Matra Transport company might have disappeared during its dry spell. Its disappearance is a little less likely today. The company has finally given itself the means to resist time. This is because its R&D component has become "relatively less important," while research is, as we know, the surest—although the most enjoyable—way for a company to go bankrupt.

"This isn't relativism any more," I grumbled, "it's mush. Okay, so you have to take your informant's frame of reference into account every time—I can see that. It's not simple; it messes up your notes. But finally you can still follow them with a good coding system, something like this: 'I myself, on June 10, 1988, heard Gontran, a non-Communist member of Fiterman's cabinet, tell me that, prior to 1985, Matra had been engaged in headlong rushes.'

"Okay, next, the actors change size—fine. If you take the trouble to figure out who's talking, whether it's the director, or his stockholders, or some underling's underling, or the janitor, you can still make it out.

"The actors' goals change—fine. All you have to do is connect each definition of a target with the starting point, t_1, and with the destination, t_2. It's complicated, but it's doable.

"If, in addition, it's not the actors themselves who define who they are or what they want but the other actors, and if you thus have to follow Matra-in-1982-for-the-RATP-in-1984, which is of course not the same thing as Matra-in-1985-for-the-RATP-in-1984, and of course not at all the same thing as Matra-in-1985-for-Bus-Division-of-the-RATP-in-1985, but not the same thing either as Matra-in-1985-for-the-Budget-of-1988 . . ., then it's already considerably less clear. We had structure before; now we suddenly have nothing but lumps.

"But if, in addition, people start mutating as the story unfolds, if there's no longer anything but the proper name that allows us to spot them, and if they go so far as to change the way they change—and all this without counting the fact that they mix up their 'ontologies,' as you put it, trusting to luck—in that case, there aren't even any more lumps; there's just mush. I knew that sociology was a soft science, but this is hypersoft, and don't try to claim that the reference mollusk is hypersoft as well, because the physics of relativity, in the last analysis, lands on its feet."

"Yes, but you understand perfectly well that they can't have any strategy because they have no interests . . ."

"Obviously, no strategy is possible any longer in such a muddle. You've completely dissolved the interests; there are neither strategists, nor uniforms, nor Ordnance Survey maps, nor drums, nor bugles."

"And that's where we land on our feet, just as surely as Einstein. That's exactly why military types have learned to draw up strategies and hierarchies, why they've invented uniforms and epaulets, why they've created the maps, why Gribeauval perfected the general staff, why military orchestras were signed up quite purposefully—precisely so a strategy could become possible no matter what. You're always going from one extreme to another. If you start from total confusion, then you understand the work that's required to clean it up."

"And that's how you hope to land on your feet?"

"Let's say that I'm letting the actors take care of straightening out the disorder they've created. You break it, you pay."

"You sound just like my father. 'You messed up your room; you go clean it up!'"

[DOCUMENT]

Transportation Ministry, Bureau of Ground Transportation. Paris, March 23, 1984. Note addressed to M. Henri, Transportation Ministry cabinet member. Subject: meeting with M. Lagardère, CEO of Matra, Inc.

The attached note presents:

—the activities of Matra's transportation division;

—the Matra group's industrial strategy in this area;

—potential directions for manufacturing;

—a report on the progress of the negotiations about the development of the Aramis CET;

—initial information about the implementation of a VAL system in Toulouse.

It appears that Matra:

—has an *ambitious* production policy that is not always very well *adapted* to the French and foreign markets;

—*needs* a short- and middle-term work plan for the development of its transportation division (hence its aggressive commercial policy in Toulouse in particular);*

—may use its financial participation in the Aramis CET as a *bargaining chip* in return for government support for a Matra project in Toulouse.

*A VAL system was under consideration in Toulouse at the time. It was inaugurated in 1994.

[DOCUMENT]

Ministry of the Economy, Finance, and the Budget: Budget Division. Letter of April 27, 1984, addressed to M. Henri, Transportation Ministry cabinet member:

Thus you would like the Ministry of the Economy to agree to the request made by the Matra organization for support for ANVAR. This would diminish Matra's contribution to financing the CET by 50 percent . . . I do not deny the existence of such hopes *[raised by Aramis]*; they have been nourished, moreover, by individuals who are technologically competent in these areas.

On this basis, an intervention by the collectivity might well have been considered legitimate.

It is nevertheless a basic principle that the manufacturer's commitment to the operation remains clearly *indicative* of his confidence in its success. That is the best *guarantee* that the State is not stepping in *just when the company itself has stopped believing in its project*.

Under these conditions, the cost sharing for the CET that was envisaged in the preparatory phases, which leaves *one quarter* of the development expenses to fall to the manufacturer, although generous, could have been looked upon *favorably*.

The request made to ANVAR, an organization financed by public funds, *completely disrupts* that financing structure by leaving Matra with only a small deductible—on the order of 12 percent—instead of a financial participation that would represent a real commitment. Thus, the request cannot be accepted.

The actors themselves are working to solve the problem raised by the relativist sociology in which they've situated one another.

They have to ensure the stability of their interpretations by establishing catwalks that allow them to go from one reference frame to another while modifying their own viewpoints as little as possible. The sociologist isn't the only one who doesn't know what Matra wants; this is also true of the minister, the minister's cabinet head, and even Lagardère himself, Matra's CEO. They, too, want to stabilize a certain interpretation of what they are and what they want. And there they are, ordering notes and questionnaires, which accumulate in a file that is soon complicated enough to require new notes, syntheses, and summaries. What does Matra want, in the end? What do we want, finally? The investigation carried out by the external observer can be distinguished only by the time frame, the budget, and the goal, from the innumerable investigations that punctuate the history of the project. The final audit merely extends the ongoing audit. To make fun of the files and the bureaucrats, to make fun of the two-page notes of synthesis and the thousand-page appendixes, is to forget the work of stabilization necessary to the interdefinition of the actors. It is to forget that the actors, large or small, are as lost in the action as the investigator is. The human sciences do not show up as the curtain falls, in order to interpret the phenomenon. They constitute the phenomenon. And the most important human sciences, always overlooked, include accounting, management, economics, the "cameral sciences" (bureau-graphy), and statistics.

In order to keep their grasp on the branches, the actors install little *valorimeters* in as many points as possible, to ensure the translation of one point of view into another. The responsible party in the Budget Office, for his part, sets up a counter powerful enough to let him measure the test of force; or else the private manufacturer takes an interest in the project and pays for, let's say, at least a quarter of Aramis; or else he asks to pay only 12 percent, and that *proves* that he's no longer interested in the project. The "deductible" is too low. As for the vice-director of the Bureau of Ground Transportation, he provided himself with a battery of somewhat more complex indicators to respond to his minister, who is going to meet the CEO: by looking at Matra's capital structure, by studying its indebtedness and its failures relative to exports, he thinks he can deduce the negotiating margin that may allow Fiterman and Lagardère to come to an agreement: Toulouse for Aramis. After all, a Marxist minister can hardly be astonished that the superstructure depends on the infrastructure of capital . . .

By multiplying the valorimeters that allow them to measure the tests in store and to prove certain states of power relations, the actors manage to achieve some notion of what they want. By doing their own economics, their own sociology, their own statistics, they do the observer's work and construct the fluctuating object that the observer will have to investigate later on. Bureaucrats are the Einsteins of society. They make incommensurable frames of reference once again commensurable and translatable. The protocol of agreement, red-penciled and ratified, starts moving again, going from one reference body to another, tracing a path along the way, a succession of fragile catwalks that make the agreement harder to break each time, because it is now weighted down with the word of the State. Officially, they're all starting to come to agreement. The relativist crisis is diminishing, the little seventeen-page memo is ready to be initialed. They come to agreement—more or less tacitly—with a compromise: the hesitations over Aramis have gone on long enough, we're going to do the CET, we're not going to spend more than 149 million francs, and we'll meet again in twenty-seven months.

"But you're exaggerating a little, Norbert. Economics does exist outside calculations, dossiers, economists, and statisticians. We're up to our ears in it. It's our world."

"Yes, as context—nothing more and nothing less. You have to follow the economization of a project just as you do its contextualization."

[INTERVIEW EXCERPTS]

Bréhier and Marey, at the RATP, are again discussing the economic studies they carried out.

M. Marey:

"We weren't worried about cars; we were worried about passenger flow. We started with a complicated network, with forks, because without that it wasn't worth doing Aramis, obviously.

"We started with around 10,000 passengers per hour during peak periods. We got up to 600 pairs of cars, assuming that each train was made up of five or six pairs of cars. All this had retroactive implications, of course.

"The rolling stock generated traffic for us, which required us to have rolling stock; then we went back to the technicians with our questions.

"Aramis has a lot of advantages because of the forks. Waiting time for a passenger goes from three minutes, with a trolley, to forty seconds.

"The problem is that everybody's seated. With Aramis you can't say, 'We'll just cram folks in. So as soon as there's even just one more passenger, you need another car.

"It isn't really an economic calculation we're making, because it's a network study. What counts is the average travel time per passenger.

"In Paris, with the orange card, users pay a standard rate, in any case, so the return on a new investment can't be calculated directly. On the other hand, we know from experience that as soon as people are offered a new system of public transportation, they give up private cars.

"So our profits come from *new* passengers; those are the only ones we can honestly count . . .

"Then there are three parts: one, the investments, the civil engineering—that's what costs the most; two, the system, mobile and fixed; next, the traffic—that is, the new passengers and the new offer of comfort; and finally the operation—that is, drivers and maintenance.

"What can be said, taking into account the narrowmindedness of the Budget Office *[Finance Ministry]*, is that all the economic studies showed that Aramis was actually pretty well situated. It was completely conceivable, even if the Budget Office was somewhat critical of the way things were calculated.

"But it was obvious that the operational costs of the rolling stock were awfully heavy. Three billion francs in investments sounded reasonable enough, but the hitch was the operating costs.

"Still, in the end, since the service was new, it was conceivable that the quality would have attracted new people."

M. Bréhier:

"We started with a socioeconomic balance sheet by valorizing the passenger, and then we looked to see whether it was tolerable from the standpoint of the financial balance sheet.

"In any case, the Budget Office finds everything intolerable; as they see it, no investment for a specific site is ever profitable. It's only if there's a political will behind a project that they decide it's tolerable. Whereas we were pushing for the new service, the time gains, the increased comfort." [no. 30]

[DOCUMENT: LETTER DATED APRIL 27, 1984]

The Budget Office response:

As for the Aramis project as a whole, and its implementation along the former Petite Ceinture in Paris, first of all, you know that the study conducted under your initiative by the RATP—and with which my ministerial department had very little to do—concluded that profitability would be marginal. The picture improves only if the calculation is broadened to take into account the time saved by passengers, a qualitatively useful notion, but a very controversial one when it comes to quantification in financial terms. And even that "socioeconomic profitability" remains modest.

Economics is not the reality principle of technology; technology has to be realized gradually, like the rest of the mechanism for which it paves the way.

Every technical project has to define a type of economic calculus that makes it more profitable. Economics is not a framework in which engineers subsequently insert themselves, one that would serve as an overall constraint; it's a simulation that mobilizes human-beings-on-paper by means of calculations. Programming flows of instructions between the Aramis vehicle and its colleagues is not so different from programming passenger flow; the two tasks require the same computers and sometimes the same equations. The engineering system endows the rolling stock with properties: one has to know where the cars are on the tracks. Similarly, the economist endows his human-beings-on-paper with properties: they will give up their cars only if they will actually save time by using Aramis. The flow of simulated rolling stock is joined, on paper, by the flow of enthusiastic passengers. So begins a calculated narrative that can hold its own against the best detective stories.

How many mobile units should there be? 200, 400, 600? All the know-how of the system's dimension-determiners is in this joint scenario that distributes roles to reasoning beings. Humans are flow; the transportation system is a network; the computer has to maximize the load factor: not too many empty cars, not too many waiting passengers, watch out for rush hour. The human being in these scenarios is an interesting fictional character, a rather new type, as idiosyncratic as Aramis' intelligent vehicle. Or rather the two together make up a new hybrid, "the potential Aramis passenger," which has surprising effects on the calculations. "Potential Aramis passengers" can't be crammed in, since they are all seated; they have to know where they want to go, since every car goes its own way; they have to give up their automobiles so the profitability calculus can become visible.

The relation between the economic calculation in camera and that in the greater Paris region is a relation to be established, to be performed, to be maintained. It is no more a given than any of the other relations. The profitability of the network and the efficiency of the rolling stock are twin notions that are negotiated and gradually realized as functions of success or failure. They follow; they do not lead. They are decided; they are not what makes it possible to decide. The fact that the Budget Office gets mad and challenges the mode of calculation does not intimidate our economists—the Budget Office would dearly love to be the reality principle for all of France. And so we have a fine scientific controversy opening up within economics to determine whether "socioeconomic profitability" is acceptable or not. If politics imposes its will on the Budget Office, then the Budget Office has to take into account the calculation of passenger time and comfort, and Aramis becomes *profitable* once again. If politics hesitates, then the Budget Office imposes its own method of calculation, and Aramis goes back into the red. Aramis will survive only if it extends the scope of its network to the point where it makes humans in Paris move and modifies the usual calculation methods of the Budget Office.

"But the consumer demand? There must be a demand. In school I did take courses in economics, after all, and there truly is such a thing as demand. If Aramis doesn't have any passengers, then the project is done for."

"Of course the demand exists—but like profitability, it was only on paper. They tested it with a psychosociological study, on a model of Aramis."

[DOCUMENT: SOFRES COMMUNICATION, DECEMBER 1983]

Perception of the Aramis System by Its Potential Audience: A Psychosociological Study.

. . . The study shows that potential users have not yet managed to position Aramis within the familiar universe of collective transportation. In addition, there is considerable reluctance to use a system in which the emotional risks and the physical disadvantages of forced proximity in a small enclosed space are much more obvious than the functional advantages of autonomy, flexibility, and speed. Aramis indeed brings to the surface the most negative aspects of public transportation (the indiscriminate mixing of people) without offering the refuge of anonymity in exchange . . .

The potential clientele is fairly open to the idea of a completely automated transportation system. In the eyes of the public, Aramis is not innovative in this area; it merely repeats a formula already tested by other modes of transportation, such as VAL.

1. The size of the cars is much more surprising, in the Aramis project, than its technological novelty.

What is novel about Aramis, in the eyes of its potential clientele, is the fact that the cars limit the number of passengers to ten and that all passengers must be seated. Aramis comes to entail a notion of personalized comfort that the public perceives as incompatible with the stereotype of public transportation, which connotes crowds, cramming, and discomfort.

2. The qualities for which Aramis is appreciated are the very features that lead to doubts about the system's efficiency in the "normal" situation of public transportation.

. . . The proper functioning of Aramis, in the public mind, presupposes not only that people are relaxed, reasonable, healthy, and disciplined, but that they agree to be "distilled," "stockpiled" in the waiting mode, in order to avoid any problems gaining access to the cars. In contrast, the subway, by virtue of its accessibility, seems to be a more reassuring system.

3. The various incidents represented and people's expressed fear of being closed in reflect the difficulty of adapting to this new mode of group travel.

. . . The small-group situation confers an exaggerated importance on interpersonal relations: the situation resembles that of a closed cell where the ten occupants of a vehicle are condemned to associate with and put up with each other in a restricted and confined space. As it is no longer possible to isolate oneself or get lost in the crowd, the slightest posture or gesture could have an unfavorable (or favorable) impact on the behavior of other passengers . . .

4. Endowing Aramis with its own special personality. Potential passengers inevitably make comparisons with the metro in order to assess the advantages and disadvantages of Aramis. In the case of the metro, indeed, individuals face a familiar universe whose rules they have definitively integrated and whose advantages and disadvantages they have accepted, whereas in the case of Aramis they face a system with which they have no experience at all . . . Aramis has to be perceived as the prelude to a new philosophy of transportation, addressed to responsible adults.

Thus, we have an inconsistent *dual image,* of which the two parts refer to incompatible systems [p. 25] . . .

The metro ends up representing a *higher psychological comfort level:* it can absorb passenger traffic more easily; potential passengers can be certain of finding a place and thus are apt to waste less time; it offers greater respect for timetables; and, last but not least, it offers the benefits of anonymity. [p. 31]

Suggested solutions:

Combat the sensation of being closed in; minimize the disadvantages of face-to-face contact; diminish the state of tension among passengers resulting from the continual coming and going; provide reassurance against the risk of violence.

The following is typical of the comments made by the passengers surveyed:

"This system," someone said, "does not accommodate the handicapped. No, it rejects them. There's no access for the handicapped, or for the blind, or for very tall people, or for luggage. The constraints are a little too constraining." "If you only have young people, okay, it'll work. If there are handicapped people with canes, then what happens?"* [p. 11]

Consumer demand and consumer interest are negotiable like everything else, and shaping them constitutes an integral part of the project.

Aramis has been under discussion for ten years, and this is the first time grass-roots customers have appeared. Petit and Bardet spoke of the French who had to be saved from pollution and the automobile; Fiterman was to speak of the "right to transportation"; the RATP's dimension-determiners spoke of passenger flow; the Matra engineers spoke of the passenger—an ergonomic and more or less idiotic being who might well panic all by himself in his car. But everyone was speaking precisely *in the name* of passengers. From the very outset, in their view, the paper passenger is enthusiastic, saved from private cars and public transportation. But will flesh-and-blood passengers subscribe to Aramis' version of them, and settle nonchalantly into the comfortable spot that the experts have spent ten years preparing for them? Do we have to wait, before raising the question, until Aramis actually exists from head to toe? No; a minimum of retroaction is required. The Aramis car has not yet been fully designed; it can be reinscribed so as to take into account the reactions of flesh-and-blood human beings who do not belong to the research bureau. The RATP orders a survey of the potential clientele. Of course, basic customers do not yet speak on their own—they are mobilized, organized, translated by psychosociologists from the SOFRES polling firm who chose a representative sample of men and women and got them to speak under controlled conditions. They're still just spokespersons, but humans are finally speaking—the report

*Here we recognize the nightmare of the transit system's creators: the little old lady or the handicapped individual with a cane. See the interview with M. Petit in Chapter 1.

even includes direct quotes! A prototype sample gets into a mockup of Aramis. A mockup public interprets the Aramis prototype.

As it happens, the representative and represented humans do not subscribe to all that has been said about them for the last ten years. The main advantage that the transportation experts are excited about—the cross between private cars and public transportation—is profoundly shocking to the man in the street. Customers constitute as disagreeable a bunch of scoundrels as readers. You make a fine book for them, one that takes them by the hand, and they call it incomprehensible; you make a lovely Aramis for them, and they call it a gadget for healthy people or for a theme park—and they claim that the metro is psychologically more reassuring! It's enough to make you tear your hair. You offer them the hybrid of dreams, futuristic transportation, and they go off and reenact Sartre's play *No Exit!* Hell is other people seated comfortably in the Aramis car! And then they reach the height of cruelty and ingratitude: here's technological prowess that has given engineers hundreds of sleepless nights for ten years, that makes them quake with fright because it may not work, and the customers take it for granted! Obviously it's automated and there are nonmaterial couplings. Big deal! They have so much confidence in the RATP's technological proficiency that they don't even notice Aramis' new sophistication.

"Nobody is enthusiastic in this business, that's for sure. Not the supply or the demand, not the Communists or the right wing, not the Ile-de-France Region or the City of Paris."

"And yet they're signing on the dotted line."

[DOCUMENT]

Protocol of agreement concerning the realization of the Center for Technological Experimentation and the approval of the Aramis transportation system.

Preamble:

The study of the Aramis transportation system by the Matra company has been receiving support for several years from the Transportation Ministry, the Region, and the RATP, which is responsible for the program's delegated contracting authority.

The development phases of this program have been prudent and progressive.

Studies and construction of components and simulation trials have been pursued vigorously and have made it possible to confirm the principal characteristics and performance capacities of the system, along with their mode of realization.

Continued development now implies that trials leading to approval of the system should be carried out on site.

This is the object of the present protocol, which involves the program of realization of the Center for Technological Experimentation (CET), qualifying trials that will define production costs; the protocol also spells out the modalities of the project's financing.

. . . . .

. . . *Article 5.* Time frames and results: the time frames indicated in months in the present protocol have as their starting point the month T_0 of notification of the RATP–Matra agreement.

The execution of the program must respect the following deadlines:

$T_0 + 27$: first balance sheet covering functional and technical performances and commitment of the Matra engineering division on system costs, accurate within 10 percent. This commitment is to be understood with reference to public agreements covering engineering and architecture.

Taking the first balance sheet into account, the parties to this agreement will study the appropriateness of launching industrialization studies for the system so as to allow Matra to commit itself as to the definitive costs.

. . . . .

Signed in Paris, July 16, 1984

Minister of Transportation	Ile-de-France Region
C. Fiterman	*M. Giraud*

Régie Autonome des Transports Parisiens — C. Quin

S.A. Matra — *J.-L. Lagardère*

"Amazing! Here it is before our very eyes: the solemn signature of a compromise embodied in a contract. The project of a protocol for the Aramis project was drawn up when the World's Fair project was going full-speed ahead. The target date, $T_0 + 27$, made it possible to test the system in 1986, in time for the production stage and for the opening of the 1989 World's Fair. Well, look at this: it's still there in the version signed in June 1984, after the World's Fair was abandoned—in June 1983. But see how vague it is? 'The parties will study the appropriateness of launching studies.'"

"How could Matra have accepted something so unattractive?"

"They tried to get more. I've gotten hold of drafts of much tougher contracts. This is the first project I've worked on, by the way, where I've gotten confidential documents in the mail . . . So Matra proposed a version that tied in the construction of the CET with the construction of the line. But the Budget Office was against it. Matra didn't insist. After twenty-seven months, the State was supposed to commit itself to financing the line.

"The twenty-seventh month remains in the protocol like a vestige of the whole misunderstanding we've been studying for the last year. Number one, it's a vestige of the World's Fair that Chirac didn't want. But look—it's also a vestige of Matra's strategy: Out of the question to build the CET without a line as a reward, yet there's no longer any question of making a firm commitment to Aramis. Number two, it's a vestige for the Region and for the public authorities, who tantalize the industrialist with the possibility of a line, yet don't commit themselves either. Finally, it's a vestige for the RATP and for the research services that hope to have at least some transportation system endorsed by the end of the twenty-seven months, but without any guarantee as to whether it will be pure research or applied.

"It's a textbook case. No one could decide whether to do Aramis or to undo it, but everybody managed to decide to do the CET, 'to see what would come of it,' and they agreed to meet two years later. To be or not to be—*that* is the question that the famous clause of the twenty-seventh month makes it possible to avoid answering before twenty-seven months are up! Everybody agrees not to make any decisions. Complete unanimity not to find out whether the set of things being agreed on is an empty set or not!"

"And, of course," I said indignantly, "you're going to stand around with your arms crossed, counting the blows?! *Suave mare magno* . . . and the whole nine yards. But don't you see that they're all getting carried away? That they're just postponing the problems to the twenty-seventh month as if they'd said a month of Sundays? 'Give me a call and we'll go out for a bite to eat,' that's what that means. Nothing is more absurd than this protocol. It refers twice to production and the commercial line after the CET, and it provides no way, none whatsoever, to produce the line. 'Maybe afterward, if we have the money, we'll look into studying the possibility of beginning to examine whether or not we might undertake studies in view of possible production. And Matra, a private industrial firm, tight with its money, comes on board, aiming at production by forced march, with no guarantees! And the Region, which has to transport passengers, puts up the financing without a squawk! And the Budget Office goes along! And the Transportation Minister agrees! It's completely irrational! If your sociology isn't capable of passing judgment on this absurdity, of condemning it, then it's your sociology that has to be condemned, no matter how 'refined' it may be."

"We aren't here to judge the actors. The actors are always right, whether they're multiplying viewpoints or cutting down on them. If we use the adjectives 'irrational' and 'absurd,' it's because we haven't made our own frames of reference supple enough. We're spell-breakers, not spell-casters. I've told you before—we do white magic, not black."

"There's a word for your type of abdication: it's *quietism!* And there's a word for your sort of total empiricism—you're a *positivist!* There's even a word for the virtue involved here: it's *cowardice!*"

Obviously, I had lost all sense of proportion. The good rapport between us had evaporated with Norbert's stubborn insistence on clinging to a relativist sociology that made him, as I saw it, relinquish the prey for a shadow.

[DOCUMENT]

Speech by Charles Fiterman, Transportation Minister:

THUS THE CENTER FOR TECHNOLOGICAL EXPERIMENTATION OF THE ARAMIS SYSTEM IS ABOUT TO SEE THE LIGHT OF DAY.

THIS IS ALL THE MORE GRATIFYING IN THAT THE PROJECT, WHICH TOOK ITS FIRST STEPS FOURTEEN YEARS AGO, GOT SOMEWHAT—LET US SAY—SLOWED DOWN ALONG THE WAY, AND THAT IT WAS ADVISABLE TO MAKE A DECISION WITHOUT FURTHER DELAY BY BRINGING TOGETHER THE NECESSARY PARTNERS.

EVERYONE HERE KNOWS THAT I HAVE WORKED STEADILY TOWARD THIS GOAL FOR THREE YEARS. AND THAT IS WHY, BEFORE ANYTHING ELSE, ON BEHALF OF THE ADMINISTRATION, I WOULD LIKE TO THANK EVERYONE WHO HAS CONTRIBUTED TO PREPARING THE WAY FOR THE STAGE THAT WE ARE ABOUT TO REACH TODAY IN THE DEVELOPMENT OF THIS SYSTEM AND ITS FULL-SCALE TRIAL. FIRST, THE MATRA COMPANY, WHICH TOOK THE INITIATIVE FOR THIS PROJECT AND UNDERTOOK THE FIRST TECHNOLOGICAL EXPERIMENTS IN THE EARLY 1970S, AND WHICH TODAY PUTS ITS IMPRESSIVE TECHNOLOGICAL COMPETENCE INTO THE "HOPPER" OF THIS PROTOCOL, ALONG WITH A FINANCIAL CONTRIBUTION; NEXT, THE RATP, WHICH HAD THE CONTRACTING AUTHORITY IN THE NAME OF THE STATE, STARTING ABOUT 1974, FOR A PROGRAM OF EXPERIMENTATION AND REFINEMENT OF THE ARAMIS SYSTEM; ESPECIALLY IN RECENT YEARS, THE RATP HAS SHOWN A VIGOROUS DETERMINATION TO MOVE THE PROGRAM AHEAD, BRINGING A FINANCIAL CONTRIBUTION ALONG WITH ITS OWN EXPERTISE.

AND FINALLY, THE ILE-DE-FRANCE REGION AND ITS PRESIDENT, M. GIRAUD, SINCE THE ARAMIS SYSTEM'S CENTER FOR EXPERIMENTATION IS TO BE LOCATED IN THE ILE-DE-FRANCE, SOUTH OF PARIS, BETWEEN BALARD PLACE AND THE BOULEVARD VICTOR, AND SINCE THE REGION WAS WILLING TO AGREE TO INCLUDE THE CONSTRUCTION OF THIS CENTER IN THE CONTRACT OF THE PLAN WORKED OUT BETWEEN

THE STATE AND THE REGION, AND TO MAKE ITS OWN SUBSTANTIAL FINANCIAL CONTRIBUTION.

THE FACT THAT SUCH DIVERSE PARTNERS, REPRESENTED HERE TODAY, HAVE COME TOGETHER TO SIGN THIS PROTOCOL ATTESTS, I BELIEVE, TO ITS IMPORTANCE AND TO ITS INTEREST.

FOR MY PART, I WOULD SIMPLY LIKE TO POINT OUT BRIEFLY THE REASONS THAT HAVE LED THE GOVERNMENT, TO THE EXTENT THAT IT IS INVOLVED, TO LEND ITS FULL SUPPORT TO THIS PROGRAM . . . THE FIRST OF THESE REASONS HAS TO DO WITH OUR POLICIES ON URBAN TRANSPORTATION . . . WE WANT TO SEE PROGRESS IN WHAT WE HAVE CALLED *THE RIGHT TO TRANSPORTATION [underlined by the minister]*—THAT IS, THE POSSIBILITY FOR EVERYONE TO HAVE ACCESS TO A PUBLIC-TRANSPORTATION SYSTEM AT REASONABLE COST TO THE INDIVIDUAL AS WELL AS TO THE LARGER COMMUNITY . . . IF WE TAKE INTO ACCOUNT BOTH THE BUDGET AND THE FUNDS EARMARKED FOR MAJOR PROJECTS, THESE LAST FEW YEARS HAVE BEEN MARKED OVERALL BY AN INCREASE OF MORE THAN 70 PERCENT IN PUBLIC FUNDS DEVOTED TO URBAN TRANSPORTATION . . .

THE ARAMIS SYSTEM IN FACT OFFERS NEW POSSIBILITIES FOR TRAVELING ON DEMAND THROUGHOUT A NETWORK, WITHOUT CHANGING TRAINS, SO IT BRINGS IMPORTANT ASSETS TO PUBLIC TRANSPORTATION AS COMPARED TO PRIVATE CARS FOR URBAN TRANSIT.

LET ME ADD THAT THE ARAMIS SYSTEM, ESPECIALLY IN TERMS OF AUTOMATION, RELIES ON STATE-OF-THE-ART TECHNOLOGIES WHICH ARE THEMSELVES OF GREAT INTEREST AND WHOSE DEVELOPMENT FITS PERFECTLY INTO THE RESEARCH-AND-DEVELOPMENT PROGRAM WE HAVE SET UP WITH THE MINISTER OF RESEARCH AND INDUSTRY . . .

WORKING THIS WAY TO EXTEND THE GAMUT OF OUR TRANSPORTATION SYSTEMS IS ALSO A WAY OF INCREASING EXPORT OPPORTUNITIES FOR OUR INDUSTRIES . . . TO THIS END, OUR TRANSPORTATION ENTERPRISES MUST HAVE ACCESS TO A SOLID INTERNAL MARKET, AND AT THE SAME TIME WE MUST MAINTAIN A CONTINUING EFFORT IN RESEARCH, INNOVATION, AND DIVERSIFICATION THAT WILL ALLOW US TO KEEP OUR PLACE IN THE FOREFRONT.

HERE ARE SEVERAL GOOD REASONS, THEN, TO SUPPORT THE ARAMIS PROJECT. ALL WE HAVE TO DO NOW IS HOPE THAT THE EXPERIMENT WILL BE FULLY CONCLUSIVE. THERE ARE GOOD REASONS TO THINK IT WILL BE, FOR THOSE OF US FAMILIAR WITH THE HIGH QUAL-

ITY OF THE INDIVIDUALS AND COMPANIES THAT ARE "HATCHING" THIS PROJECT. IT IS TRUE, OF COURSE, THAT EVERY HUMAN ENDEAVOR INCLUDES AN ELEMENT OF RISK. WE ARE FULLY AWARE OF THIS RISK, AND WE ACCEPT IT, KNOWING THAT TO RISK NOTHING IS TO GAIN NOTHING—SO LONG AS WE PROCEED IN SUCH A WAY THAT THE RISK IS CAREFULLY CALCULATED AND ACCOMPANIED BY ATTENTIVE EFFORT.

SO GO TO IT—YOU HAVE MY VERY BEST WISHES FOR YOUR SUCCESS!

"It's written in capital letters so he can read it out loud. It still doesn't count as enthusiasm," Norbert continued without giving an inch. It 'got somewhat—*let us say*—slowed down along the way'; it 'attests, *I believe,* to its importance'; 'there are *good* reasons to think so'; 'to risk nothing is to gain nothing, *so long as* the risk is calculated." He can tell that there are glitches somewhere. But what's most poignant is the fact that, four days later, the Communist ministers left the administration.* Fiterman, proud of his balance sheet, is leaving us. Bye bye, Fiterman. Aramis has to go on without him."

"And then without Matra, and without the RATP, and without the Region, and without the City of Paris, and without the Budget Office."

"Nothing gets done with the Budget Office; everyone agrees on that. If there were nobody but the Budget people, we'd still be traveling by ox-cart."

The interpretations offered by the relativist actors are *performatives.* They prove themselves by transforming the world in conformity with their perspective on the world. By stabilizing their interpretation, the actors

*After supporting the Mitterand administration for three years, the Communist Party decided to break its alliance with the Socialists, whom the Communists accused of not living up to their revolutionary ideals . . . Following party instructions, the four Communist ministers resigned, albeit somewhat reluctantly, so as not to compromise the party in a "culture of government."

end up creating a world-for-others that strongly resembles an absolute world with fixed reference points. When the office of the radical Paris mayor decided, around 1880, to construct the Paris metro at last, it was producing a highly unfavorable interpretation of the great railroad companies: "They're wild capitalists, human beasts; it's out of the question to turn them loose in Paris and allow them to interconnect their stations." But the interpretation of elected officials is a fickle thing. What one election does, another can undo. How can the major companies be forever discouraged from invading Paris? Bingo: by writing the unfavorable interpretation into law; by signing a protocol. But a protocol can be denounced. A nonaggression pact can turn back into a scrap of paper. Elected officials then find a much better solution: they cast their unfavorable interpretation in bronze, iron, cement, steel. How to do things with words? By turning them into performatives. So they dig the tunnels of their new metro, giving them dimensions such that the smallest of the capitalist wagons will be unable to penetrate them,* even if by chance the radical mayoralty should lose future elections. The big bad wolf can blow down the first little pig's house of straw, and by blowing a little harder he can knock down the second little pig's house of wood, but he'll blow in vain on the third little pig's house of brick. No matter how relativist the engineers in the pay of the big companies may be, they find themselves facing thousands of tons of stone which impose on them a "relatively absolute" interpretation. The best proof of this bizarre mix of relativism and absolutism is provided some seventy years later when the nationalized SNCF and the national RATP with which it is fraternally united want to interconnect at last. Having become a nice little pig, the SNCF had a lot of trouble getting rid of the thousands of tons of stone piled up by the third little pig between itself and its ancestor, the big bad wolf.† What required just a few strokes of the pen in 1890 required ten years of work starting in 1970. The radicals' unfavorable

*On this episode, see Maurice Daumas, ed., *Analyse historique de l'évolution des transports en commun dans la région parisienne, 1855–1930* (Paris: Editions du CNRS, 1977). They had even thought of changing the width of the tracks, a still more radical proposal to which the military authorities in charge of national defense objected in the name of the country's higher interest: munitions cars had to be able to pass from one network to the other.

†The traces of this relativism can still be seen in every labor strike at the Gare du Nord, and they can still be heard every time the subway trains pass from the SNCF version of electric power to the RATP version.

opinion of the capitalists had "performed" reality in a more durable matter than the flighty history of France required.

In July 1984 the minister has not yet reached the stone house, but he thinks he's gotten past the paper one. The "good reasons" to support Aramis strike him as solid. And they must be solid indeed, since they have scarcely changed since Bardet's and Petit's time. Even if the Communist minister raises the level of debate still higher, right up to the "right to transportation," a new human right, it's still a question of competing with private cars thanks to "on-demand service" and the absence of "transfers"—whereas the line projected for the Petite Ceinture has been transformed into an omnibus with three or four forks! It's still a question of developing research at a propitious moment—even though the goal of the CET is an industrial goal of qualification and homologation presupposing that the research as such has already been completed. Finally, it's a question of helping companies to export by offering them an internal market—even though no commitment at all has been made regarding the industrial orders that will follow the CET. Each of these good reasons to sign on to Aramis at last, three and a half years after the project was started up once again, cuts through doubts that appear in the solemn discourse, even if the powers that be "are fully aware" of those doubts and "accept" them. From relative absolutism to absolute relativism, there is no more than a nuance, a partition of paper, wood, or stone. Let's hurry to inscribe in the nature of things the interpretation we have produced of the nature of things. Time is short. Especially if the Communists are in a shaky position and are going to withdraw a few days later into the refuge of their working-class fortress. Let's sign fast, or all is lost. Let's at least get ourselves the shelter of a sheet of paper. It's better than nothing. A few days more, and it really would have been all over for Aramis.

[INTERVIEW EXCERPTS]

M. Gontran, technical adviser to the cabinet, is once again explaining how people at the ministry felt when the agreement was signed:

"Nobody took responsibility for stopping . . .

"The problem, in France, is that the operating agencies combine the commercial function with the R&D function. They're both judge and interested party, and they lack the means for *radical* criticism of a project."

"But at the time of the final decision in 1984, what was the motivation exactly?"

"Two things got mixed up together. I think there was a confusion of plans.

"There were never any major concerns about making a Javel-Bercy connection—that is, a mini-VAL. People told themselves, 'There's a guard-rail'; it was reassuring. 'We're innovating with an innovative gimmick that hasn't been proven, but if something goes wrong we can always land on our feet with a mini-VAL.'

"This brings us back to Matra's industrial logic, which could always make the most of an inferior version . . .

"Matra wanted a mini-VAL line after the CET, which was also the goal of the Transportation Ministry, but because of their priorities they never managed to make the project concrete.

"In relation to a *rational process* of more or less logical sequential development, the various stages kept getting mixed up. They moved faster to stop the operation than to start it; a lot faster, for no reason." [no. 42]

"He's the one who's talking about irrational processes and illogical behavior," I said furiously, "not me. He's the one who's accusing the operating agency of being both judge and interested party, and that's a serious charge. Everybody who signs the protocol to build Aramis hopes to do something other than Aramis. That's really not normal, is it? They spend 150 million francs in order to wait, simply because no one has the guts to say they shouldn't spend the money! The whole thing is cobbled together in defiance of common sense. Why should we be the only ones who don't have the right to pass judgment?"

"Go on, pass judgment if you feel like it," Norbert replied. "Go right ahead, lay the blame! Pick out a head to roll for the vengeful populace. Denounce the profiteers, the ignoramuses, the incompetents."

"I'm not necessarily looking for a guilty party," I said more prudently, incapable as I was of selecting a head to chop off right in the middle of the bicentennial of the French Revolution.

"So what are you looking for? Are you going to accuse the social system? Capitalism? Napoleonic France? Sinful man, while you're at it?

An accusation that's so watered down it blames the whole wide world is even more futile than one that picks a scapegoat."

"But you're still not going to claim, are you," I went on in a more conciliatory tone, "that the protocol signed is the best possible compromise? Next to you, Pangloss would come off as a pessimist. In any case, didn't you commit yourself to finding a guilty party, a cause for the death of Aramis?"

"If Aramis is dead, which remains to be proved; if it ever lived, which also remains to be proved; and if it was viable, which has not yet been demonstrated . . . We don't have a cause. On the other hand, we have circumstances by the bucketload. Installation in Paris; choice of the Petite Ceinture; support from the RATP; abandonment by Matra, which had other things on its plate after 1985; separation of the mobile unit, which was being perfected, from the system, which was being deferred . . ."

"But those aren't explanations. They could all turn out in Aramis' favor."

"Precisely. They're symmetrical. Failure and success are explained in the same terms. Every one of Aramis' peculiarities would have disappeared if the system were working today. They strike you as scandalous and irrational only because the system fell on its face. If you were looking at VAL or the Airbus, you'd have exactly the same shambles."*

As I felt myself seething with rage once again in the face of these "scientific" explanations capable of explaining away both hot and cold simultaneously, and because I didn't want to cause the tension to rise any higher, I dropped the subject.

"You're indignant," continued Norbert, with a coolness that struck me as feigned, "because you want to blame someone; you blame me for indifference because I blame no one. But no rule has been violated in this Aramis story. No one has behaved badly. No one would have known how to behave better. You wouldn't have known how to do any

*See the documents about VAL's political and administrative history collected in Arthur Notebart, *Le Livre blanc du métro* (Lille: Communauté Urbaine de Lille, 1983).

better. Everything happens in defiance of common sense, but there is no common sense for innovations, since they happen, they begin, they invent common sense, the right direction, the correct procedure."

"Still, a mistake must have been made, either in 1987, in stopping, or in 1984, in starting up."

There are two major sociologies: one is classical, the other relativist (or rather, relationist).

Only the second allows us to follow the realization or the derealization of technological projects. Classical sociology knows more than the "actors"; it sees right through them to the social structure or the destiny of which they are the patients. It can judge their behavior because it has fixed reference frames with respect to which the patients behave in a pathological fashion. It has its ether. There are norms, and thus there are deviations with respect to the norm; there are reasons, and thus there is irrationality; there is logic, thus there is illogicality; there is common sense, and thus perverted senses; there are norms, and thus there are abnormality and anomie. Classical sociology can comment on what the patients say because it possesses metalanguage, while they have only language. "Forgive them, Father, for they know not what they do." For classical sociology, the actors are *informants*. Classical sociology explains what has happened, blames, denounces, rectifies. It offers lessons. Its judgments are above the fray; they are scientific; an abyss separates them from the interested interpretation of the patients obliged to perform the reality that the sociologists analyze. Classical sociology knows what constitutes society, knows the rules and laws of the social context within which the patients cannot help but be inserted. For classical sociology, there are classes, socioprofessional categories, fields, roles, cultures, structures, interests, consensuses, and goals. Classical sociology is at home in social physics, and it chooses an ontology that allows it to define once and for all the nature of the power relations and interests pursued by the strategy of social groups. Finally, aware of the countless contradictions entailed by its own existence situated at once above the fray and in the middle of it, at once inside society and outside it, classical sociology multiplies its methodological precautions, its hermeneutic circles, its retroactions, its marks of modesty. Classical sociology is epistemological; it talks and talks and goes on talking. Obliged to reassure itself continually as to its own scientificity, it can be recognized by its jargon. For classical sociology, the world is an asylum of fools and traitors, of pretenders, guilty consciences, and half-educated types. In this asylum, the sociologist is the director, the only one who has the right to go outside.

You can study anything with classical sociology—anything except the sciences and the technologies, anything except projects. They go too fast. They become too soft or too hard. For sociology, they are like an extended Michelson-Morley experiment. It is impossible to detect the ether in relation to which they displace themselves. To study them, you have to move from classical sociology to relativist sociology, and see in the former only a particular case, an approximation, a valuable one to be sure, but only when nothing more is moving, when projects have become *objects,* institutions. Relativist sociology has no fixed reference frames, and consequently no metalanguage. It expects the actors to understand what they are and what it is. It does not know what society is composed of, and that is why it goes off to learn from others, from those who are constructing society. It adds its own interpretations to those of the actors whose fate it shares, often less felicitously than they. It seeks, too, to perform reality in order to keep its own version of the facts stabilized a little longer, and it confronts, fraternally, the contrary opinion of those it is studying. It has no strings to its bow but theirs, and it does not allow itself to throw "science" onto the scale in order to unbalance the equality between itself and its brothers with whom it is conversing. Without any knowledge other than what it gets from them, it is free at least from the crushing responsibility of being more scientific than the actors. No guilty conscience, no epistemology encumber it, and thus no jargon. For relativist sociology, indeed, everything is grace.

They're all gathered around the hors d'oeuvres. I'm becoming irresistible, they say. I move on; they move on to the act. They sign the acts. The die is cast. The checks are signed. The word is given. For fifteen years they've been meeting about me, for fifteen years they've been speaking about me in a vacuum. Have they finally decided to make me? Now they're reassuring each other about my feasibility. They're all afraid, I sense that, but seeing themselves all together around me makes them feel braver. I am lying in the middle of the large circle they are making around me, these ministers, cabinet heads, mu-

nicipal officials, public employees, and engineers large and small. At the center of this big circle, I am the deep well into which they are tossing their wishes, their hopes, and their curses. Blessed, cursed. Loved, hated. Indifferent, passionate. Plural, singular, masculine, feminine, neutral. I am waiting for them all to grant me being. What is a self? The intersection of all the sets of acts carried out in its name. But is that intersection full or empty? I exist if they agree, I die if they quarrel. But if they agree on a misunderstanding, how could I manage to exist? How can I keep the Communists in the government? How can I keep the drivers' unions under control? How can I diminish the threat of wholesale automation in Paris while being the first wholly automated metro in Paris? How can I hold onto the region and its enormous, irritable, impatient crowds who have to be transported throughout the southern suburbs of Paris and who see me merely as a metro like any other? How can I hold on to Matra, who loves only my older brother, VAL, and who loves me only as another VAL? How, in spite of everything, can I hold on to the tuned-in engineers of the world of transportation who love me because I'm crazy, bold, and beautiful, but only on the condition that I transport no one, and especially on the condition that I do not resemble VAL, which is already passé, old hat. Some love me assembled and rigid like a metro; others love me scattered, dispersed, experimental like a research project. How can I keep about me, in agreement about me, those who think that I am an infeasible system, that I am stillborn, that I am an idea from the past, an idea from the Sixties, that I am, consubstantially, a failure, but that from my erratic behavior one may draw interesting results, perhaps at least a variable-reluctance motor, guidance software? . . . As if I were a motor, a casing, a program! A heap of separate parts! As if I could exist without being assembled. I'm quite willing to satisfy any one of them; I'd like to satisfy them all. But they'll have to come to agreement about me. How can I become a being, an object, a thing—finally a self, yes, a full set, saturated with being—without them, without their agreement, without their coming to terms (since I myself am made from them, flesh of their flesh, a rib extracted from theirs), without their acknowledgment that I am transports, displacements of human beings? How can I interest them all in me when they all love me differently? I can give them only what they have given me. I can hold them assembled together only if they keep me assembled. The "I" that humans receive at birth—that is precisely what has to be created for me. I am in a prebirth state. I do not yet have a body. The dismemberment humans encounter in the tomb is my condition even before the cradle. Reverse conception, which exists in fact only in Erewhon. The breath of life to which I aspire in order to make of my scattered members and my whitened bones a being that is not of reason—my soul—awaits your agreement, O you hors-d'oeuvres-eaters, who all agree today to defer my genesis until later. Indifferent to what you love. Rubicon-crossers who set up camp in the middle of the ford. Human beings contemptuous of things and thus contemptuous of yourselves.

6 ARAMIS AT THE CET STAGE: WILL IT KEEP ITS PROMISES?

"I don't get it, Norbert, I don't see what you're doing," I confessed, feeling both angry and disappointed. "We know that the key to the puzzle lies in the phase between 1981 and 1984. We're finding splendid guilty parties: the three-and-a-half-year delay, the fact that the project was maintained unchanged even though the World's Fair had been abandoned, the complete ambiguity of the decisions, the screening role played by the technocrats, the postponement of all the problems to the CET stage, the signing of the protocol by a government minister who stepped down three days later, and—poof—you wave your wand and pronounce everyone innocent; you decree that there's no basis for bringing suit; you send off all the accused parties with kind words, you tell them they're great; you claim they wouldn't have done anything differently if the project had been, in the last analysis, a complete success."

(Silence on the part of my beloved professor.)

"In my opinion, you're simply too chicken to blame people who have ways to get even. If you want to know what I think, your 'hyperrefined sociology' is a pretty flimsy affair."

"Except, in spite of everything, it isn't a political project," Norbert replied, more uncomfortable than I expected. "It's a technological project. And after all, it's a fact that we haven't yet seen any crucial defects on the side of the people involved, or in the way things were organized, in the decisionmaking process—in the social arena, as you'd put it. If Aramis existed, you wouldn't find anything to criticize. There still isn't any proof, at this point, that the thing couldn't work. I have no choice but to find the people innocent. I have absolutely no evidence to the contrary."

"But what's left for us to look at, then? The guilty party isn't at the end, it isn't at the beginning, it isn't at Orly, it isn't before 1981, it isn't after 1981. Where is it? Or isn't there a guilty party at all?"

"Well, yes, there has to be; a mistake was made, since Aramis doesn't exist today, yet it did start up again in 1984. There was at least one decision too many; you're right about that. Either the decision to terminate it in 1987, or else the decision to start it up again in 1984."

"So what's left for us to look for? That's what I'm asking you."

"There's Phase 3B, there's the CET, there's the technology; and since the technology really takes off during this phase, we have to go that route. Follow the actors: that is the Law and the Prophets."

"But you've been holding technology blameless from the beginning, *a priori;* you've been saying that we have to assume it was feasible and well conceived! That I don't have the right to say that this bloody mess was defective from birth, deformed, monstrous—that it's the very prototype of a false innovation. You've made me read page after page on the subject."

"No, that was about technology that was separated, nonhuman, inert, autonomous. I'm not budging an inch on that. You can't blame technology. But there must be something else in the technological aspects. It's like in *The Mystery of the Yellow Room.* If the guilty party can't be found anywhere else, it has to be there."

"But where is 'there'?"

"You're driving me crazy . . . I don't know yet. I'm looking. If I knew, I wouldn't go to the trouble of doing all these interviews. We have to go into the technical details, that's all I know; they're responsible for holding this whole shambles together. We've got a good grasp of the politics of the interphase, but a whole lot of technological work was going on for a while there. They spent six or seven months reconstituting their teams, and then let them go again soon afterward. We'd better back up a little."

"It's the 'stop-and-go' disease, Norbert—that much we know."

"Yes, but they're reformulating the project, so we're going to find out whether the technology is capable of coping with the politics. The Aramis we knew prior to 1981 has to be redesigned to cope with the World's Fair."

"But I've studied the technological documentation, as I was told, and I've summarized the interviews you sent me to. The engineers were very pleased with that phase; everything went very well. So the problem isn't there. The hidden staircase, as you put it, can't be in this phase."

"It has to be, or else everything happens in the CET stage."

[SUMMARY OF A GROUP INTERVIEW]

On the boulevard Victor, in the now-abandoned offices that had belonged to the RATP team, the engineers in charge of the project summarize the development of Phase 3B.

"Matra had to work out quite a few new things during that phase. If Aramis can only take curves greater than ten meters, then it's hard to fit it in between the World's Fair pavilions or in lightly populated urban areas—or heavily populated ones. If Aramis is hard to implement, then the flexibility argument disappears.

"Since highly flexible implementation is the main argument in Aramis' favor in the eyes of the municipal authorities, especially because of Montpellier, it's essential to keep the curve radius as tight as possible. So Matra has to come up with a curve radius of ten meters.

"During this period, the old Orly site was reopened. They constructed a curve with a ten-meter radius to see if the vehicle could take it without slipping and without pulling the contact shoes away from the power source. At the price of shifting the location of the shoes and widening the guide-rails from 1,400 to 1,440 millimeters, Aramis became capable of taking curves with a ten-meter radius.

"The other major problem was that of availability. If Aramis cars are to be in circulation, they have to be able to run ten hours a day without breaking down. Now the only way to achieve greater availability is to double or triple the rolling stock. But each vehicle has only ten seats. So any improvement in reliability will be paid for by only ten passengers.

"So we had to increase the absolute reliability of the equipment. Now this is only marginally possible—everything always breaks down, even when military technology is used, even in aeronautics, where Matra was on pretty familiar ground. So in every car all the equipment had to be duplicated. But that was impossible because of the cost.

"So the size of the cars had to be increased; that way the costs of automation would be divided up among more than ten people. However, you can't lengthen the vehicle or widen it without losing the ten-meter curve radius that we'd just taken so much trouble to get.

"So Matra had to find a solution that would keep the radius and improve availability without increasing costs. They came up with *paired cars,* two vehicles physically attached together—a solution imported from VAL. Each pair of cars now holds ten passengers and the probability that the identical equipment on the two singletons will break down decreases. At the price of a physical connection between pairs of cars, the reliability, the price, and the flexibility are maintained." [no. 7; see Photos 11–13]

Mechanisms cope with the contradictions of humans.

Pairs of cars transcribe, or take upon themselves, or accept, or transfer, or take the place of, the technological impasse (chips are fallible, alas), the economic limitations (the vehicles must not cost as much as satellites), the contradictions of the system (the mobile units cannot simultaneously grow longer and turn readily). The connecting bar that attaches the singletons two by two *shoulders* the responsibility of simultaneously holding together the mobile units and the contradictions of the engineers' erratic demands, exactly the way the nonmaterial couplings that still connect the pairs of cars *shoulder* the responsibility of resolving the contradictory dreams of those same engineers—behaving as if they were automobiles, churning up a massive flow of cars guided like trains. But since the two dreams coexist, the result is material coupling between cars in each pair, and computerized coupling between pairs!

Although charged by humanists with the sin of being "simply" efficient, "purely" functional, "strictly" material, "totally" devoid of goals, mechanisms nevertheless absorb our compromises, our desires, our spirit, and our morality—and silence them. They are the scapegoats of a new religion of Silence, as complex and pious as our religion of Speech. What exegesis will have to be invented to provide commentary on the Silence of machines? What secular history will ever be able to narrate the transcription of words into the silence of automatons?

Beyond our infinite respect for the deciphering of Scripture, we need to have infinite respect for the deciphering of *inscriptions*. To propose the description of a technological mechanism is to extract from it precisely the *script* that the engineers had transcribed in the mechanisms and the automatisms of humans or nonhumans.* It is to retrace the path of incarnation in the other direction. It is to rewrite in words and arguments what has become, what might have become, thanks to the intermediary of mechanisms, a mute function. The physical link between cars, the calculated link between pairs of cars, and the little shock absorbers that authorize shocks: here is the morality of things.

"Hmm," I said skeptically, "we're doing theology now, are we? Considering where we've got to, actually, we may as well start lighting candles; we're just drifting. Anyway, from day one you've been explaining that we have to hold onto the humans and the nonhumans. I don't see what's new in Phase 3B."

"But it's the Achilles' heel of the whole Aramis project! We hadn't yet grappled with the technology. It was still only a project, and projects are words, plans, signs, that sort of thing, whereas now we're going right to the object, and we're still dealing with the social arena, the social bond, attachments and values, but now they've been altered, transformed. The question now is to find out what quantity of the social element Aramis can absorb, transform, displace, by getting more complicated, by folding itself up tighter. If it can hold its whole contradictory environment together, then it will exist."

*Madeleine Akrich, "The De-Scription of Technical Objects," in Wiebe Bijker and John Law, eds., *Shaping Technology / Building Society: Studies in Sociological Change* (Cambridge, Mass.: MIT Press, 1992), pp. 205–224.

[SUMMARY OF A GROUP INTERVIEW]

The same speakers:

"If the Aramis cars feel each other out by way of ultrasound detectors, they run the risk of bumping into each other or staying too far apart, because ultrasound echoes are scattered in tunnels and on curves. We'd have to separate the trains and slow down the cars when they merge and demerge. But that would disrupt the rhythm and reduce the speed, and Aramis' advantages, its train configuration, would disappear. So Matra has to find something else to maintain the performance levels.

"All guided-transportation systems, trains and metros, are based on the principle of *fixed sectors*. The track is divided up into sections and no vehicle is authorized to enter one section before the vehicle ahead of it has left—that is, before the red light of each section has turned green. Trains and subways are equipped with a simple injunction: never cross a sector's red light, whether the light is interpreted by a human being or by the motor directly.

"But if the train is going ten kilometers an hour, it can brake faster than when it's doing sixty, so it can follow the train ahead more closely without increasing the danger. However, the sector doesn't authorize this, because it remains unaware of this margin for negotiation. It maintains a constant distance between trains, no matter what their speed.

"It's like a cop who takes away all the flexibility from trains and subways in exchange for a considerable margin of security.* A procedure like this isn't appropriate for Aramis, since the car has to be able to connect up with the cars that precede it and form a train with nonmaterial couplings. So it has to *penetrate* the sector ahead by feeling its way. But if it's authorized to go into the safety sector, then its means of recognition have to be absolutely failsafe. Otherwise, if it's too forceful, it'll crash into the other cars, or, if it's too timid, it'll stay so far away that it'll slow down the flow of the vehicles behind.

"Matra comes up with a radical solution, one already considered for VAL but then abandoned because of its complexity. It involves doing without the sacrosanct fixed sectors, in exchange for increased intelligence on the part of the cars. The Aramis vehicle is endowed with the means to negotiate its own speed and safe distance *on its own,* no longer by means of fixed sectors

*Even SACEM—a model for assisted driving, recently installed on Line A of the RER, which makes it possible to bring trains closer together than the human drivers present in the cabin would dare to do—reduces the intervals only by dividing up the fixed sectors into smaller units. It does not do away with fixed sectors.

installed on the ground, but by means of reference points inscribed on the track that allow it to calculate its own speed and that of the cars ahead and to *decide,* in view of its own program of action, whether it will authorize itself—or forbid itself—to move closer to a colleague. This is the adjustable mobile sector (CMD), Aramis' major innovation: it transfers the notion of sector from the ground to the mobile unit and makes it flexible instead of rigid. As each vehicle moves forward it projects its own *danger zones,* both ahead and behind—forbidden zones whose *dimensions vary* according to the speed of the vehicle and the speed of the adjacent cars.

"But then how can safety be maintained? When the cars are far enough apart, the CMD is enough to discipline them. When they approach zones of convergence or divergence, the CMD will be supplemented by a way of feeling blindly ahead; ultrasound detection will be abandoned as too unreliable, and will be replaced by a system of detection by means of hyperfrequencies, which are much less sensitive to interference from tunnels or time. Ultrasound echoes will be used only for short distances, tens of centimeters, when the cars are moving together in a train." [no. 7]

A technical project always gets more complicated because the engineers want to reinscribe in it what threatens to interrupt its course.

The scriptor, or the engineer who had delegated the driver's job to a closet in VAL, was nevertheless as mistrustful of the closet as of the driver; both are fallible. That is why he had maintained the principle of fixed sectors on the track. The program of action—"Keep your distance, don't run into each other," which the people responsible for automobiles have so much trouble inscribing in the conditioned reflexes of drivers—was transcribed by the scriptor into a different one: "Don't enter sector *n* before sector $N+1$ is free." And this program was itself retranslated for human drivers into yet another one: "If relay switch in up position, then power supply to motor on; if relay switch in down position, then power supply off." And this one was also transcribed into another, an electrical diagram or software program.

But the cascade of such translations, while maintaining safety, abandons flexibility, and thus ceases to authorize Aramis. Either the principle of the flexibly configured train has to be scrapped, or else flexibility has to be *reinscribed,* but with no loss of security. Here in a nutshell is all the engineering, all the ingenuity, all the ingeniousness of engineers, those unsung writers. For they never consider people or things; they consider only *competences* that can be reshuffled more or less freely. The capability of maintaining order, for example, was once transferred by the inventors of railroads to the ground, which had been marked off at regular intervals into fixed sectors. Why not redistribute this capability? Let's leave to the ground, to the track, the task of simply noting where we are. Let's displace *onto the mobile unit* the task of negotiating braking distance. Let's thus reincorporate into the confines of the vehicle a part of its environment. Let's thereby offer the system a blend of flexibility and safety that is unknown to automobile drivers and that no labor union would agree to give to the agents of the RATP, since the briefest moment of inattention would lead to catastrophe.

The CMD is the most beautiful invention in the world. It reinscribes in the car what until now has formed the *environment* of the mobile unit circulating in fixed sectors. Aramis is growing more complicated, the script is folded and refolded; but the characterization is deepening, the character is taking on depth, subtlety, and body, since it is becoming both flexible and sure of itself! Technology is sociology extended by other means.

"But you've been saying that all along."

"I hadn't understood myself. By other means: there's the heart of the matter—by other means. You have to respect otherness, the others, and mediation, the means. Here's the key to the enigma. The CMD isn't social, it isn't political; it's calculations. But it isn't technological, either. It isn't an object. This Phase 3B is important. It's the first time they've redone Aramis, quickly, from top to bottom, instead of simply implementing it and improving it bit by bit. So it was doable! Aramis could exist. All it had to do was absorb all the rest of the contradictions. All it had to do was remain an agency of translation, of reinscription, a whirlwind, a soul."

My mentor's behavior worried me a little. He insisted on our doing "lab work" on the most ordinary machines. "We have to find the mistake," he would mutter. "A crime has been committed against Aramis, and we now know where to locate it: in the reinscription, the folding." Carried away with this logic, he routinely thanked the automatic ticket machines at highway toll booths. He queried automatic tellers at banks about communication problems. He had long conversations with electric staplers. He noted the degree of politeness, laziness, violence, or nastiness of all the automatic door openers he came across, going so far as to tip them, which usually left them quite indifferent. He couldn't buckle up a seat belt without looking into its stiffness, flexibility, or looseness, undoing the springs in order to see where that morality of webbing and clasps could be coming from. One day he undertook a complete interview with a "sleeping policeman," a speed bump, on the pretext that this peace officer was more faithful, more serious, more intrinsically moral than his own nephew, though the latter was a precinct captain. Brushing my worries aside, he claimed that his own director used to converse frequently with coquilles Saint-Jacques in the Saint-Brieuc bay. "So you see?"* He demanded that I respect my alarm clock on the pretext that the moral contract I had signed with myself—and that I tried to forget as soon as I sank into my dreams—was faithfully preserved by the mechanism and punctually recalled in the form of the alarm bell. He wanted me to get my electric food processor to admit what it took me for—an idiot, I discovered, in dismantling the thing, since it was impossible to make the blade go round without having carefully closed the cover; this alone, through the intermediary of three little sensors, authorized the motor to start. At the hotel where we stayed in Lille, he even went so far as to get the porter to acknowledge the moral of the French hotel-key story. Why are those keys so heavy? Because they remind clients that they have to comply with the program known as "Bring your keys back to the desk, please," a program that tourists, as careless as they are undisciplined, keep on forgetting, according to the bellhop, but that they are reminded of by the weight of the key in a pocket. He wanted the bellhop to share his admiration for this moral law finally ballasted with lead. I was mortified.

*M. Callon, "Some Elements of a Sociology of Translation: Domestication of the Scallops and the Fishermen of St. Brieux Bay," in J. Law, ed., *Power, Action and Belief: A New Sociology of Knowledge?* (London: Methuen, 1986), pp. 196–229.

"We've found the hidden staircase," sighed Norbert without noticing my embarrassment. "This is how the guilty party got in. It has to be this way. Either Aramis transforms itself to hold on to its environment, and it gains in existence, or else its environment gets away from it, it no longer copes, no longer reabsorbs what's out there, and it loses in existence."

[SUMMARY OF A GROUP INTERVIEW]

The same speakers:

"So we've maintained a curve radius of less than ten meters, we've shifted from single cars to pairs, we've invented the CMD: to all this, Phase 3B adds several other transformations involving noise pollution and the air gap of the variable-reluctance motor that had to be fiddled with a bit.

"But the most important choice for the fate of Aramis is the shift to microprocessors in response to the need for increased intelligence in the mobile unit. The entire set of calculations can no longer be hardwired, as was traditionally the case in rail systems; there are too many of them. It would take kilometers of wire, and it would take forever to check all the connections. The engineers have to resign themselves to programming—softwiring—the set of functions, and to achieving safety by verifying previously encoded calculations. Rather than being intrinsic, security now becomes probabilistic. It is less certain, since the programming *language* is what now ensures protection rather than the relay or transistor board *circuits*."

Every technology may be a project, an object, or an exchanger.

We have been mistaken. Up to now, we have believed in the existence of objects. But there are no objects, except when things go wrong and they die or rust. Holding on to the adjustable mobile sector is feasible provided that one doesn't try to maintain safety conditions. However, it is impossible not to maintain safety conditions, since the entire legal system of the world of transportation depends on it. Should the CMD be abandoned? Should safety be abandoned? No, let's burden the new microprocessors with the weight of our dilemma. Let's entrust our calculations to them! Let's reinscribe in them the entire set of action programs that we can no longer take care of by legal, social, or traditional means. The microprocessors become the center of the new Aramis, the principal object of the future Center for Technological Experimentation. They hold everything. Five years ago they didn't exist; now they are making Aramis possible at last.

Where is this being, the microprocessor, to be situated? On the side of human beings? No, since humans have delegated, transcribed, inscribed their qualities into nonhumans. On the side of nonhumans, then? Not there either. If the object were lying among nonhumans alone, it would immediately become a bag of parts, a heap of pins, a pile of silicon, an old-fashioned object. Thus, the object, the real thing, the thing that acts, exists only provided that it *holds humans and nonhumans together, continuously.* Slightly out of phase, it resides neither in the social element (it is made up of chips and hinges, shock absorbers and pairs of subway cars) nor in technologies (it is made up of passions, transported people, money, Communist ministers, and software). On the one hand, it can be said to hold people together, but on the other hand it is people who hold it together.

Give me the state of things, and I'll tell you what people can do—this is how technologism talks. Give me the state of human beings, and I'll tell you how they will form things—this is the watchword of sociology. But both of these maxims are inapplicable! For the thing we are looking for is not a human thing, nor is it an inhuman thing. It offers, rather, a continuous passage, a commerce, an *interchange,* between what humans inscribe in it and what it prescribes to humans. It translates the one into the other. This thing is the nonhuman version of people, it is the human version of things, twice displaced. What should it be called? Neither object nor subject. An instituted object, quasi-object, quasi-subject, a thing that possesses body and soul indissolubly. The soul of machines constitutes the social element. The body of the social element is constituted by machines.*

A soul? A body? Naked men? Isolated automatisms? Improbable. Of course, there are versions and interpretations—social, sociologizing, subjective, spiritual—of the objects so instituted; people are said to exist and to live in society and to speak. Of course, there are technological, technicist, objectivizing, material interpretations of these same objects: they are said to consist of raw, inanimate material and to dominate, or to be

*Bruno Latour, "Ethnography of a 'High-Tech' Case: About Aramis," in P. Lemonnier, ed., *Technological Choices: Transformation in Material Cultures since the Neolithic* (London: Routledge, 1993), pp. 372–398.

dominated by, people. But there is never any question of the quasi-object, which bears so little resemblance to people and so little to machines. We have been mistaken. What we had called the "technological object" is what lies on the garbage heap, in the scrap pile, abandoned by people and by other projects.

The only object we have met in this story is the Aramis car in Matra's foyer, unless we count the CET site, which has become a haven for graffiti artists and the homeless [see Photo 2 and Photo 20].

[DOCUMENT: REPORT ON THE END OF PHASE 3B, AUGUST 2, 1983]

Conclusion: the work undertaken in phase 3B indicates that the critical problems have been satisfactorily resolved and that full construction of the center for technological experimentation can be undertaken.

"One more phase declared innocent!" I exclaimed. "You see? It's gone as smoothly as clockwork. All we have left is the CET, Phase 4; that's our last hope."

"If they've managed to maintain the object in this state of trouble, of turbulence, negotiation, exchange, reinscription, in the CET phase, then they're bound to succeed; they're going to turn Aramis into an animate body. Aramis is going to live for real," Norbert cried out, quite forgetting the sad end to this story, although he had known it all along.

"So we tackle the last phase?"

"Yes, of course! Let's head for the boulevard Victor."

"I'm finally going to see something solid instead of reading documents and technical notes. I'm finally going to do my job as an engineer."

"You'd better count on being disappointed. There are still a lot of documents to examine at the CET" [see Photos 9 to 18].

"But the CET is much more complex than anything we've studied up to now. How are we going to find our way around?"

"The actors have the same problem we do," Norbert replied, unruffled. "They, too, have to find their way around. So they must have invented a solution. If it were really complex, they'd flounder."

[INTERVIEW EXCERPTS]

A Matra technician in charge of planning:

"The classification plan that makes it possible to code each document is itself a twenty-eight-page document, and since its importance for management is enormous, every new updating is *initialed* by everybody responsible for the project."

"It's a reflexive document."

"Reflexive, I don't know; in any case it provides a cell in its grid—code number 184,100—designed to guarantee the numbering and the management of the documentation. From the CET (no. 100,000), a continuous path lets us go forward or backward—for example, to the platinum plating (no. 124,112) of the ultrasound sensors (no. 124,000) on the automated devices (no. 120,000). You see, with this numbering system, we don't get lost; it's manageable, its workable, and it actually has worked very well" *[see Figure 9].*

It is in the detours that we recognize a technological act; this has been true since the dawn of time.

And it is in the number of detours that we recognize a project's degree of complexity. A monkey wants to get a banana that is hanging from a branch. The monkey is readily identified as a creature of desire. If he stops staring at the fruit and explores all the sticks lying around his cage, he's called a first-order technician, since he has suspended the first program in order to use a second one. If no stick is long enough, but if he takes the time to attach two pieces of wood together, transforming a short stick into a long one, he is said to be a second-order technician, because the detour itself has been suspended by a third. If the chimp were as well organized as Matra, he would code his flow chart as in Figure 10.

Technologists seemingly follow infinitely more complicated programs than those tested in cages, laboratories, or classes by their psychologist or primatologist colleagues. However, these programs cannot be much more

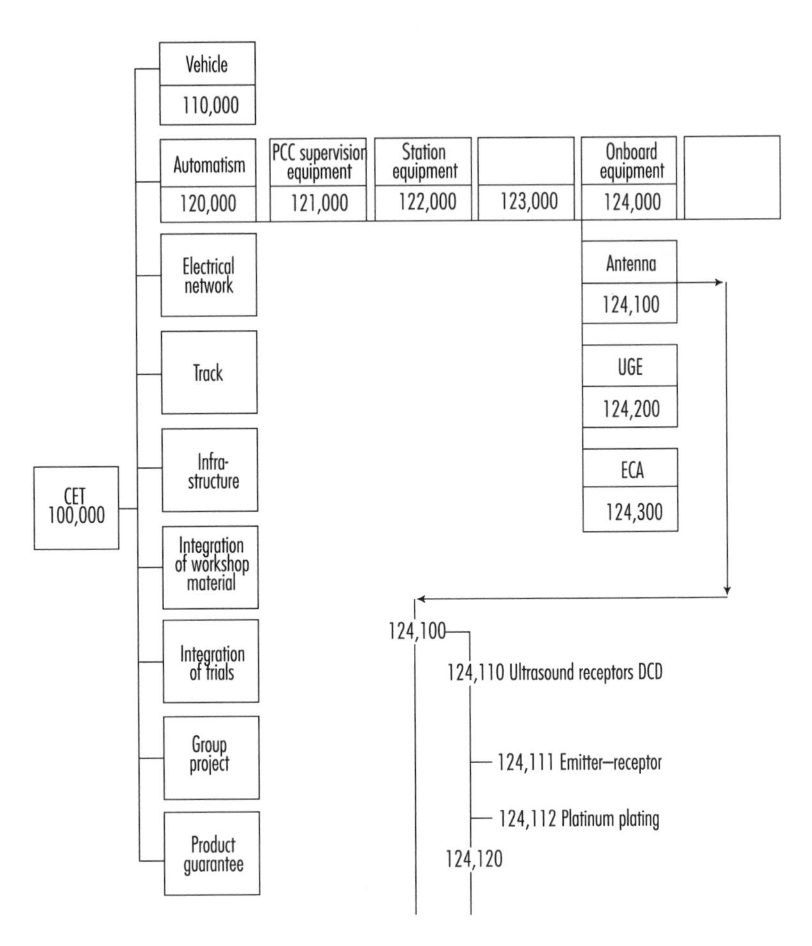

Figure 9.

complicated, otherwise the ones technologists study would become hopelessly embroiled, just as technologists themselves would. Once the ultrasound sensor has been set up according to instruction 124,110, what comes next? It is impossible for humans as well as for monkeys, for engineers as well as for ordinary mortals, to answer that question without other technologies for management, visualization, coding, and recording that make it possible to pass the rest of the task on to one's neighbor. No labor without division of labor, and no division of labor without management and coding, without files and flow charts. Our bureaucracy—so widely scorned—is our second brain, as indispensable as the first. The engineer

Figure 10.

in charge of task 124,110 has just a few subprograms to run through before she reaches her goal. Aramis, for her, consists in a circumscribed task whose ins and outs disappear above and below. It turns out to be ensconced inside a black box, but a black box with plugs, since the next one "up"—124,100—can take the sensor as a whole, and install it in the antenna—and the next one "down"—121,111—can also take the sensor as a whole, as a landscape, as a horizon, within which it installs its own box.

The circumscribing, the coding, and the visualization of the division of tasks allows a piling-up of Russian dolls that increases the complication of the whole, yet the technological object, in the eyes of a given observer, never increases in complication. Wherever the observer is located, she will never see anything more challenging than that flow chart whose content will sometimes be Aramis as a whole—no. 100,000—and sometimes the platinum plating on the casing of the ultrasound sensor—no. 124,112. The entire technological wizardry lies in the impenetrable partitions and in the pegs that make it possible to hook one's task to a neighbor's. The paradox of a technological object with millions of instructions is that it is, from the standpoint of the division of labor, a fractal object that is equally simple at every point, and that the whole looks nevertheless like a Leviathan that infinitely surpasses human measure.

I was only half reassured by the arguments Norbert was dredging up out of administrative sociology, for if we are to believe our informants, the whole CET had drifted a good bit.

[INTERVIEW EXCERPTS]

M. Frèque, project head, is speaking at Matra headquarters:

"The problem with Aramis is that the railway system is safe, but heavy and hard for Aramis' light cars to use. We really needed something like automobiles, but automobiles aren't reliable stock; they're nothing like railroads in that respect—nothing like VAL, even—and VAL still used a lot of elements from the standard subway. Obviously aeronautics and space materials are reliable and lightweight, but they're extremely expensive!

"So you see the problem? We had to invent everything.

"Everything had to be done, or done over. Since everything had to be redone—making automobile equipment reliable, lowering the cost of the Mirage III pumps (you see how far we went?), lightening the components of the subway—we were in a state of complete uncertainty, at the beginning, as to costs.

"The components didn't exist, and that made it very hard to price them. Builders laughed in our faces when we went to ask them for estimates on the price of new materials, in small quantities, to be delivered, possibly, in 1992–93!" [no. 6]

M. Laffitte, RATP engineer in charge of automation, standing in front of the Aramis prototype that was hung in the maintenance workshop after the project was halted:

"Aramis was a hybrid. That's what Frèque always said: 'Take automobile-quality material and make it as solid as railway and as sophisticated as aviation.'

"Besides, you have to reckon with technological development that is differential, uneven.

"The motor worked out pretty well. The mechanical side went all right; so did the hydraulic side. Electricity—there we know our way around. But electronics, especially microprocessors—that was another story . . .

"It was going off in all directions. We really had the impression that they weren't in control of their programmers. Each one had his own budget, went as far as he could; but there was a whole team of young guys, very smart but not very disciplined, most of them working as subcontractors. The sense of teamwork was lost.

"The alarm bells kept going off, but—let's put it this way—they sounded off internally."

M. Parlat:
"They said to our faces, 'Stop screwing around.'"
M. Laffitte:
"Stop ringing the bells, you mean."
M. Parlat:
"Right! *[Laughter]* That's it, stop ringing the bells!" [no. 3]

The work of folding in technological mechanisms can go from complication to complexity.

This is because technological detours go from zero to infinity according to whether the translation goes through *intermediaries* or through *mediators*. The VAL user takes VAL without even thinking about it. As soon as he has incorporated this slight detour into his accounting and his habits, he is almost incapable of recalling whether or not he has "taken" the metro or not—a few minuscule details, a few trivial incidents, a face, a poster will perhaps allow him to tell the difference. "Taking VAL" is a subprogram that has become nothing but an intermediary between an actor and his goal. It is, literally, a means. From the observer's viewpoint, it will scarcely be possible to detect any break in continuity between the detour and the return to the main task. The part—the subprogram—is inferior to the whole.*

Aramis' project head hasn't reached this point yet. Fortunately, some of the intermediaries occupy the precise place expected of them in the planning, without interfering with neighboring tasks or slowing them down. For complication is just the opposite of complexity: a complicated task is one made up of many steps, each one of which is simple; a complex task, as the name indicates, is one that simultaneously embraces a large number of variables, none of which can be identified separately. A computer—at least when it is working—is complicated, but it is never complex.† An ordinary conversation is often complex, but it may well not be complicated. In Paris, the subtask "Make concrete and deliver it to the boulevard Victor on time to pour the slab for the building" shouldn't pose any particular

*See Bruno Latour, "On Technical Mediation," *Common Knowledge* 3, no. 2 (Fall 1994): 29–64.

†Tracy Kidder, *The Soul of a New Machine* (London: Allen Lane, 1981).

problem. This task, for the project head, is what taking VAL is, for the average user. Entire lines of the flow chart consist of such intermediaries—piled up, folded in, folded over, implicating each other, black boxes stacked up one upon another. The building lines (no. 150,000) or the electricity network (130,000), or even the track (140,000) offer no surprises. No imps are about to pop out of these black boxes. Or at least it is easy to make them go back in. Even if calamities threaten to emerge, Pandora's box can be closed up again quickly: the concrete arrives on time, the track is ready at the right time, and the electricity lights up at the expected frequency—there is nothing mysterious in this fidelity to allies, for they have been disciplined, *shaped,* over a century or two; they have become reliable technologies, disciplined resources.

But unfortunately for the project head, Aramis requires allies that are not so disciplined—chips that skitter about in corners, programmers who are as immature as their technologies. We move, then, from the lovely flow chart to the schema of translation, from black boxes to gray boxes, from division of labor to undivided chaos. The main program is interrupted. And as the worst can always be counted on, the whole stack of subprograms can revolt one after the other. Instead of finally bringing Aramis into existence, on the boulevard Victor, the project's participants found themselves in the research situation they thought they had left behind.

Pandora's box cracks open and calamities emerge one after another. Is program B a means? or C? or D? Nobody knows. Does program B—or C or D—count at all, or has it become a definitive obstacle? Nobody knows. Each one has become a mediator that now has to be reckoned with, for it transforms the goals and redefines the hierarchy between main and subordinate, goal and means. What was complicated has become complex. Between 0 and 1, the observer must become as patient as Cantor and count numbers to infinity. The part has become superior to the whole. Every engineer grappling with new projects has experienced this mathematico-ontological drama, and many, like Cantor, have been scarred by it.

Stabilized state	*Instability*
Complicated	Complex
Intermediaries	Mediators

Fabrication	Research
Means	Ends or means
Counts as 0 or 1	Counts as 0 or infinity
Part inferior to the whole	Part equal to, inferior to, or superior to the whole
Flow chart	Labyrinth

We were now advancing with infinite care, for without warning any subprogram of the CET might become a trap door into a dungeon where no one would ever find us. While the engineers were plunging into the mud of the 1984–1987 construction site, we ourselves were plunging deeper and deeper, throughout 1988, looking for the subprogram where Aramis had been lost. I did not yet know that my mentor was going to get lost along the way before I did.

[INTERVIEW EXCERPTS]

"There are a lot of people crammed in here," *said M. Laffitte, unfolding the program of the Aramis system computers.*

"And those six computers were just the ones each car carried on board. Just look at the book of functional descriptions: 167 pages!

"This volume is the part that's easiest to digest. It's the best synthesis; it's what makes it possible to draw up the specifications. It's obviously very complex.

"As for us, we didn't have an overall view of the thing. We were only contractors. Normally we're the contracting authority and we know what's going on in detail, but in this case, with Aramis, we didn't.

"Knowledge was somewhat opaque.

"Even Matra didn't have a very clear picture. Matra subcontracted a lot of work to software outfits.

"In the end, I think everybody was treading water. Besides, software is abstract; nobody knows how to debug it very well. Wires, cables, relays, intrinsic security—all that, we're familiar with. We've been perfecting procedures for checking circuits for a hundred years; it's a matter of materials. But software? It's not concrete. We were all pretty much groping in the dark.

"What I think is that the Matra engineers were treading water so they wouldn't go under." [no. 16]

Thanks to computers, we now know that there are only differences of degree between matter and texts.

We knew perfectly well that a black box is never really obscure but that it is always covered over with signs. We knew that the engineers had to organize their tasks and learn to manage the division of their labor by means of millions of dossiers, contracts, and plans, so that things wouldn't all be done in a slapdash manner. Nothing has a bigger appetite for paper than a technology of steel and motor oil. We were well aware that they had to draw, calculate, anticipate the shape of each piece on plans and blueprints; we knew that every machine is first of all a text, a drawing, a calculation, and an argument. No machine without a design *department;* no huge machine without a huge design department.* We were well aware that mechanisms are saturated with instructions for using them, with technical notices and maintenance diagrams that make it possible to read them like a book. Every machine is scarified, as it were, by a library of traces and schemas. We were well aware that thousands of sightings, "looks," sensors, feelers, signals, alarm bells make it possible to transcribe by sight on a *control panel* what the mechanism seals up. No machine without its control panel.

But we still thought, in spite of everything, that the agents mobilized by machines eluded forms and programs. We thought there was a frontier beyond which one really moved into matter, that inert and cold stuff, functional and soulless, which earned the admiration of materialists and the scorn of the humanists. But no, calculators continue to accumulate layers of forms and diagrams, adding them to other forms and other masks, half spirit and half matter, half imprint and half text, without our ever crossing the famous barrier between sign and thing, between spirit and matter. Thanks to microprocessors, we know that "processes" proliferate constantly at all levels, from the infinitely large—organizations—to the infinitely small—electrons. In fact, ever since a literary happy few started talking

*See P. J. Booker, *A History of Engineering Drawing* (London: Northgate, 1979), and the fascinating testimonies offered in E. Robbins, ed., *Why Architects Draw* (Cambridge, Mass.: MIT Press, 1994).

about "textual machines" in connection with novels, it has been perfectly natural for machines to become texts written by novelists who are as brilliant as they are anonymous. Programs are written, chips are engraved like etchings or photographed like plans. Yet they do what they say? Yes, of course, for all of them—texts and things—act. They are programs of action whose scriptor may delegate their realization to electrons, or signs, or habits, or neurons. But then is there no longer any difference between humans and nonhumans? No, but there is no difference between the spirit of machines and their matter, either; they are souls through and through, and the gain makes up for the loss. The disorder that is wiped away on one side by describing the tasks meticulously in neat logical trees turns up again on the other side, among the programmers, who are having as much fun as a barrel of monkeys, shooting themselves in the foot, dividing up tasks according to procedures that can't be described, for their part, in neat logical trees.

If we were going to have to add the proliferation of chips and computer program lines to the multiplication of flow charts and the shambles of the muddy construction site on the boulevard Victor, I really could no longer see how we were going to get rid of all these hordes so as to do their sociology and pronounce anything like a clear judgment on that chaos.

"That's precisely the problem, the key to the puzzle. Were they able to keep body and soul together? That's really the issue. If so, Aramis lives. But if body and soul, the social and the technological, are separated, then it will die," said Norbert. I was not reassured.

[INTERVIEW EXCERPTS]

M. Alexandre, head project engineer under Frèque, is speaking at Matra headquarters:

"When we started up the CET, everything was new. As far as all the standard areas went, such as the tracks or the rolling stock, we stuck to the contractual timetable quite closely; we weren't off by more than a few weeks or months.

"As for the automation, I agree with Laffitte: we got a little behind, and we noticed it right away.

"As early as the second meeting with the RATP, we had lost ground. We underestimated the complexity of the automation involved.

"Besides, it was difficult to find teams at T_0; we had trouble sticking to the plan.

"The shift to digital technology made things more complicated. With VAL, too, there was some of that, but less. VAL has intrinsic security: all breakdowns are analyzed, and as soon as there's a breakdown, VAL defaults into security mode. With VAL, there's a speed program inscribed on the track. VAL reads it, but *doesn't know* what it's reading.

"With Aramis it was much more complex.

"With the CMD, the functional needs are different. The cars are *localized* on the track. They've committed to memory the invariants of the line* and at the end of the line they get *assignments,* or else by way of the UGT, the traffic-control unit, *they're told,* 'You're the head of the train,' so they take all that into account, they know it, and they *make a decision;* plus they may also incorporate safety constraints.

"On Aramis, there are permissive and nonpermissive zones. If the passenger panics in a permissive zone, the train stops and the door is released. In a nonpermissive zone, a tunnel for example, the train first goes back to a station.

"This is different from VAL, because the Aramis car can be precisely located, whereas with VAL, we simply know that it's in such-and-such a sector, but where exactly we can't say.

"That's what's so clever about the CMD. The vehicle calculates the position of the one ahead; it calculates its own anticollision distance. If it hasn't been told to join up with a train, it stops.

"If it has been told to join up with a train, it has *the right* to approach and to bump lightly into the car ahead—that's what the famous shock absorbers are for. It's a physical constraint—it's not *a right* the car gets. There's no way, with the refresh time allowed, to get a linkup without a bump, but it's a very small jolt at three meters per second.

"The shock absorber was invented because they couldn't get total anticollision. It's inevitable; you can't even demonstrate with calculations that it could be avoided! And anyway, the passenger can put up with it; the only alternative

*For example: "A curve, slow down to 20 kilometers per hour; a straight line, accelerate to 30; a merge zone, watch out."

is to slow down the linkups, or to spread them over hundreds of meters. In short, we were obliged to come to terms with these little bumps [see Photo 19].

"So, in fact, we had to hire young engineers and call on software companies. We didn't have the manpower in house to do SACEM and Aramis both.

"SACEM is another story. Okay, the principle is simple. You've got a functional string coded in sixteen bits without security coding, plus another string in which the information is coded. There's no redundancy; we don't double up on computers, we just redo all the calculations in parallel, and if the two calculations don't agree, the security system requires the functional system to shut down. The security system has priority.* This is the house option; it's open to discussion. In our view, relying on three computers and a *majority* vote doesn't provide any more security than coding. But that's a problem of religion; I'm not competent to judge. If they'd told me to make it redundant, I would have opened my umbrella and said, 'Why not?'" [no. 25]

Technological mechanisms are not anthropomorphs any more than humans are technomorphs.

Humans and nonhumans take on form by redistributing the competences and performances of the multitude of actors that they hold on to and that hold on to them. The form of Aramis' shock absorbers is a compromise between what Aramis can know—the speed and position of mobile units—and what humans can stand without discomfort—shocks of less than three meters per second. Let us note that here humans are being treated as objects that do or do not resist shocks, while nonhumans are granted knowledge, rights, a vote, and even refreshments. The shock absorber *absorbs* a certain definition of what can be done by the humans and nonhumans that it bumps and that it is charged to link gently together.

Anthropomorphism purports to establish a list of the capabilities that define humans and that it can then project through metaphors onto other beings—whales, gorillas, robots, a Macintosh, an Aramis, chips, or bugs. The word anthropomorphism always implies that such a projection remains

*The functional system sends orders for movement. The security system checks to see if these orders are correct by comparing them to the orders it has in its memory. This amounts to doing the calculations twice—the first time openly, the second time in coded form—and comparing the results. If there is the slightest difference between the two strings of calculations, the functional order is not carried out.

inappropriate, as if it were clear to everyone that the actants on which feelings are projected were actually acting in terms of different competences. If we say that whales are "touching," that a gorilla is "macho," that robots are "intelligent," that Macintosh computers are "user friendly," that Aramis has "the right" to bump, that chips have "a majority vote," and that bugs are "bastards," we are still supposing that "in reality," of course, all this fauna remains brute and completely devoid of human feelings. Now how could one describe what they truly are, independently of any "projection"? By using another list taken from a different repertory that is projected surreptitiously onto the actants. For example, technomorphisms: the whale is an "automaton," a simple "animal-machine"; the robot, too, is merely a simple machine. Man himself, after all, far from having feelings to project, is only a biochemical automaton.

We give the impression, then, not that there are two lists, one of human capabilities and one of mechanical capabilities, but that legitimate reductionism has taken the place of inappropriate anthropomorphism. Underneath projections of feelings, in this view, there is matter. Some even go so far as to claim that projections should be forbidden, and that only designations should be allowed. No more metaphors. Figurative meanings go into the wastebasket; let's keep only proper meanings.

But what can be said of the following projection: "The chips are bugged"? Here is a zoomorphism—bugs—projected onto a technology.* Or this one: "The gorilla is obeying a simple stimulus-response"? Here a technobiomorphism—the creation of neurologists—is reprojected onto an animal. Or this one: "Chips are only electron trajectories"? Here we have a phusimorphism projected onto a technology. But what can we say about the following sentence: "Aramis can bump if it wants to; it's not a right that it is given"? Is this an anthropomorphism? Do rights come from human feelings? From nature—phusimorphism? From gods—theomorphism? And what do you know, here we are in the middle of a philosophical quarrel. The engineers are right in the middle of it when they hesitate between their vehicle's "you can" and "you must," and when they decide that human security "has priority" over "the functional," in calculations and decisions. What? Could they be moralists, theologians, jurists, these poor engineers who are said to be pigheaded calculators (a zoomorphism and a techno-

*In French, computer chips are designated by a different zoomorphism, *puces* ("fleas").

morphism)? Let us say that, in their workshop as everywhere else, *form* is in question, that there is never any projection onto real behavior, that the capabilities to be distributed form an open and potentially infinite list, and that it is better to speak of *(x)-morphism* instead of becoming indignant when humans are treated as nonhumans or vice versa. The human form is as unknown to us as the nonhuman.

"You're going to make some friends among the human-rights crowd, Norbert, not to mention the technologists . . . And if we don't know what form humans take, or things either, how are you going to come up with your final diagnosis? And we've just about run through all the suspect phases. The CET is shaky, but all CETs are shaky, as I understand it; that's what they're set up for, as centers for technological experimentation."

"It all boils down to knowing whether Aramis, as they say, keeps its promises or not. If it keeps them, it continues to exchange properties and competences, and it comes into existence. If it's shaky, that doesn't matter, so long as they continue to exchange forms and in that way give body to their dreams."

We attacked the last phase once again in order to find out how well Aramis had kept its promises. "There's the beast all laid out," we thought as we unfolded the huge sheets of the volumes of technological specifications, page after page.

"They've never showed this," Norbert said, whistling in admiration. "It's Victor Frankenstein's workshop, it's the making of the creature itself! And look how Shelley was wrong again. Crowds, more crowds. And in her novel she describes only one tête-à-tête between Victor and his disgusting anthropoid! A technology isn't one single character; it's a city, it's a collective, it's countless. All of Germany and Switzerland together would have been needed to keep Victor's awkwardly stitched-together creature in existence. Look at all these people! And this is just an attempt at a prototype. It's only the sketch of the head of a line. It's missing the arms, the trunk, and the feet *[Figure 11]*."

I was less intrigued than Norbert by this plunge into the CET's

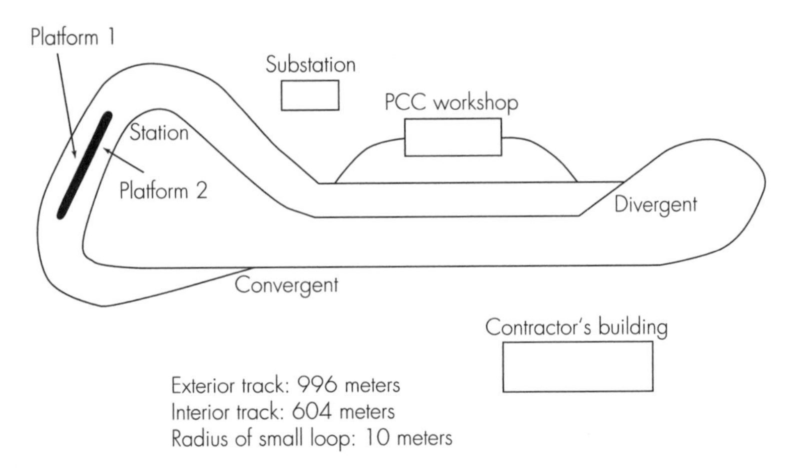

Figure 11.

technological documentation, because it was finally beginning to resemble what I had learned in school.

"And look," he added, "no matter what page we unfold, it's as complicated, as populated as the others, no matter what the scale. Up here it's the boulevard Victor. And down below, it's the inside of a single pair of cars. Each time, it's a squeeze. Six complete computers connected by a communications loop where as many things are circulating as on the loop up there, on scale one. A single pair of cars is as complicated as the system as a whole. And now each packet of bits on the internal circuit is a pair to be identified, with which to encode, to locate. Here's the title we ought to use in our report: *Crowds Press onto the Boulevard Victor-Frankenstein!*"

I understood the problems now, but I also saw why the engineers were in their element.

"And even so," I explained gleefully, "we're not taking into account the flow matrices, with 600 pairs of cars for 14,000 passengers, that the real system would have had if it had existed. The whole CET is only one little pair—nothing at all in comparison with the entire Petite Ceinture system" *[Figure 12]*.

"What crowds, what crowds!" Norbert exclaimed naïvely. And now that they've been taken on, they mustn't be abandoned. Was Shelley

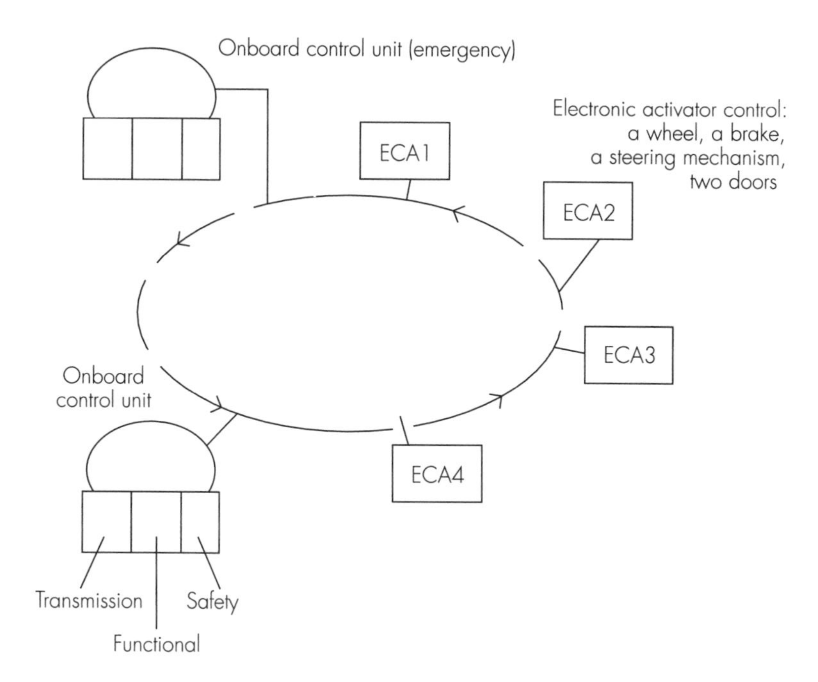

Figure 12.

completely wrong, then? Did she lie about the way the monster was made, as well as about Victor's crime? There's the hidden staircase," my mentor added.

"If I've understood correctly, Matra chose to organize four levels, four ranks, four classes."

"You see, what did I tell you? Doesn't Matra have its sociology, its political philosophy, its *Weltanschauung?*"

"If you insist," and I explained to him what I had understood.

"Level I: the pairs of cars, which not only serve to move passengers about by replacing people's physical strength by a motor, but on which automatons also embark. These automatons are the UGEs (onboard control units), which replace the driver.

"Level II: the track, which serves not only to transmit a powerful electric current—high tension to make the variable-reluctance motors work—but also a weak current, which makes it possible to create a platform for communicating with the UGE.

"Level III: the fixed equipment, each unit taking care of a sector and a station. This equipment serves as intermediary between the mobile units and the command post, and has to sort things out locally in order to solve most problems without making everything go back to the top, which would go crazy if that happened."

"They're *missi dominici.*"

"If you insist on your political metaphors. Level IV: finally the head, the central command post, which sees, understands, feels, decides, acts, orders, and manages the entire flow, but which would crash if all the information from each mobile unit were to come back to the center. Here's what we've got," I said, unfolding a new plate *[Figure 13].*

"Terrific!" Norbert said. "There are human beings only at the two ends: in the pairs of cars, there are passengers carried away with enthusiasm; in the Central Command Post (PCC), there are operators who steer and drive. Between the two, in the mobile units, on the tracks, along the sectors, everything is done by nonhumans—but by tens of millions of them. And these nonhumans have names and capabilities; they're human parts. Now don't tell me that's not a fine piece of anthropomorphism!"

"What's funny is that the PCC can't be a dictatorship, and here the political metaphor holds up, because as I understand it, there's no way for a human being both to control 660 pairs of cars and to steer them by remote control. It would be like driving 660 cars at a time without an accident. In any event, even if the PCC were made up of Formula One drivers like Alain Prost, you can't transmit enough information in both directions fast enough to react at 50 kilometers an hour and at a distance of 10 centimeters. You need a minimum of democracy—that is, delegation of tasks: the mobile units have to fend for themselves, in part."

"Ah! You see! So this time we're doing politics again for real—politics in things. I'm playing the dictator, you're playing the democrat," said Norbert, whose choice did not surprise me.

And we replayed the software in order to test Aramis' viability.

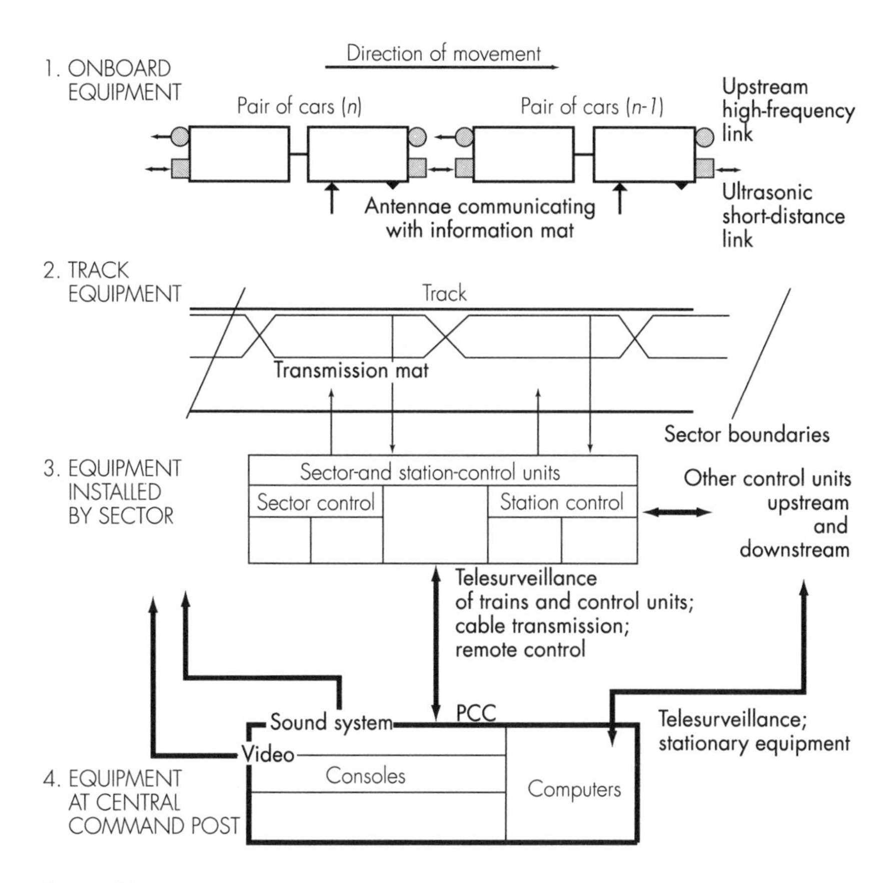

Figure 13.

[DOCUMENT]

Organization of the "automatisms" substation:

3.2.1. The Onboard Equipment

—An outboard control unit (UGE) provides for the functions of safety, driving, and supervision within each car.

ME: "I go too fast—you can't send me enough information fast enough. You have to let me have an autonomous personality. I have to drive myself."

HIM: "Well then, take care of things yourself, if you're so smart.

But since you're not human, I'd be surprised if you were capable of doing much."

A pair of cars has two redundant UGEs, one per car.

ME: "If one of them breaks down, the other one takes over. I am a pair of nonhumans. Pilot and copilot. I'm not asking you for very much; that's the condition for being autonomous."

HIM: "Okay, but two nonhumans don't yet make one human."

—An electronic interface between the UGE and the subsystems to be controlled in the car (doors, brakes, traction, steering) called an ECA [*électronique de commande des actuateurs* (electronic activator control)].

ME: "No one will open the doors; it's too dangerous—someone nonhuman has to open them. No one will put on the brakes—someone nonhuman has to put them on and make sure they're working in synch with the car's speed, so they won't take hold too violently. No one will check to make sure the motor follows suit and idles—someone nonhuman will have to check it. No one will be there to say whether or not the onboard steering is on the correct side—someone nonhuman will have to say it. And all those someones—me, them, him, it, someone—have to calculate fast and have to be checked themselves by the UGE, which is the onboard conscience."

HIM: "Conscience! You're not afraid of a thing! *Computo ergo sum.*"

ME: "Yes, but that ought to please you, Norbert, for I'm becoming someone."

HIM: "Watch out! It's us, it's I who make you become someone."

ME: "It comes down to the same thing. In the end, I am someone, since I am the origin of my own actions. You let me be the origin, granted; but then it's no longer you, it's me. The ties are cut, the delegation is irreversible."

HIM: "We'll see how long you hold up on your own . . . and how many Matra engineers you'll need to set you up as the origin of your own actions."

—Three antennae controlling the various ground contacts: one digital-emission antenna; one phonic-emission antenna; one general-reception antenna.

HIM: "He's arguing over the senses I'm supposed to endow him with! Shelley didn't anticipate this in her novel. She gave him eyes and ears, arms and hands—that's all; she didn't have much imagination. She didn't have any antennae, or ground platforms, or activators, or variable-reluctance motors. And what about consciousness? Victor bestows it on his monster with a single gesture, without meeting any resistance, without realizing that consciousness is gradual, that it can come in bits and pieces, as redundancy, self-diagnosis, feedback. There's nothing to be gotten from literature, clearly," said Norbert.

As a good engineer, that's what I had thought all along, but I'd kept it to myself.

ME: "I don't want to be a monad set up with enough refinements to carry out my program with no communication with the outside world. I don't want to be either programmed or solipsistic. I would become delocalized. I would no longer know if I were head of the train or the caboose; I wouldn't know what the others were doing, or even who I was. If I were a ballistic missile during a nuclear war, every effort would be made to isolate me and to create within me the environment of total war that would suffice to guide me. But I am not—I must not be—an engine of death."

—The phonic wheels, mounted on the front wheels, and the sensors of the "traction" subsystem, mounted on the rear wheels, supply information about the distance traversed and, through processing, about absolute speed and acceleration.

ME: "I'm struggling against myself. That's because I'm subject to the principle of relativity. I don't know where I am, or even if I'm moving. I have to be given the means to register my displacement, to keep a trace of it. I need absolute reference points."

The phonic wheels of each right front wheel are said to be security wheels (they contain coded information).

ME: "If I begin to exist autonomously, I have to be suspicious of myself; the phonic wheel might slip, or get stuck, and then it would send false information which, once processed, would make the UGE imagine that I'm going faster or slower, or that I'm not where I think I am. All these decisions could lead to catastrophe. So the crucial information contained in the phonic wheel has to be protected against betrayal by being encrypted, as in wartime. I protect myself by a double service of intelligence and counterintelligence."

HIM: "And all this so you won't have to obey my orders."

ME: "You've missed the whole point, Norbert. You can't give me enough orders, not fast enough, not in time. You have to let me handle things on my own."

HIM: "And just how are you going to do that, now that you've gotten so smart?"

This information is a safety measure owing to the physical encoding of the cogged track of the phonic wheel and of a supplementary sensor, C3. Every appearance of the timer signal, C2, is associated with binary information supplied by C2; the sequence of these bits has to correspond to a code.

Encoding makes it possible to detect any reading defect (an unseen parasitic cog); this leads to an exit from the code, thus to an alarm. Encoding makes it possible to get a secure reading of the direction of travel, supplied by the direction of reading of the coded sequences. [p. 40]

HIM: "I understand why science fiction writers take the easy way out," exclaimed Norbert, somewhat exasperated, finally, that I had become so independent and that some of the technological details he didn't get came easily to me. "They provide their own fantastic beings, *dei ex machinarum,* whereas we see the gods, the little gods, come out of the machinations. It's a whole lot harder."

An average onboard car-to-car hyperfrequency link (from 0 to 30 meters) allows the transmission of digital information from one pair of cars to the next (anticollision messages, alarms, governance instructions for coordinated travel).

ME: "Hey, listen, now I need relations with others. I want to be able to govern them, or to be governed by them. In spite of their good will and their three antennae, the control units set up on the edge of the tracks aren't capable of transmitting enough information fast enough for us mobile units to handle things at high speeds. So we have to set up direct relationships between ourselves; we have to be able to tell each other things like, 'You're going too fast, you're going to run into me, I'm your boss, you have to go at my speed, watch out, I'm braking, careful, I'm in alert status, pay attention.'"

HIM: "But there'll be nothing left for me to do. You're stripping me even of the job of setting up interpersonal relationships. What about the master-slave dialectic? You're going to take that over, too, I suppose?"

I didn't know what that was, but I didn't care, because little by little I was becoming the Aramis mobile unit. I understood how it worked and, like it, I was taking on confidence and personality. I no longer wanted to be a lowly student constantly lorded over by his mentor-master. Norbert had been living on my labor for a year, and I no longer needed his gratitude. I was the one, now, who was dictating my own technological choices. I had fought hard to win the right to recognize myself as autonomous. I was no longer afraid.

—A short-distance car-to-car ultrasound link (from 0 to 5 meters) carries out a direct measurement of relative distance between pairs of cars and also allows the recapture of information that facilitates coordinated travel.

ME: "In spite of their good will and their antennae, my hyperfrequency links aren't capable of transmitting enough information about the small accelerations of the cars when we have to link up without bumping into each other too hard. The car that follows has to be suspicious of the one that precedes. Cars thus have to be endowed with a means for measuring the distance from one to the next, directly and locally. Instead of saying slowly, 'I am so-and-so, I am moving at such-and-such a speed, my acceleration is delta such-and-such,' the car tells itself in the third person singular, 'So-and-so is at such-and-such a distance from me.' It says this by cries, echoes, and reception of the echo."

HIM: "As always," Norbert murmured, more and more annoyed by my new self-confidence, "as soon as you have autonomy, as soon as you have consciousness, you get chitchat. This Aramis car is a real magpie."

—An interphone mechanism provides for bilateral liaisons between users and operators of the PCC (interphone mode) and allows messages to be broadcast in cars (sonorization mode).

HIM: "Okay, that's enough. I'm taking back control over all your idiocies in the direct mode, manually, by voice and by sight. The fun's over. The passengers and operators have to be able to connect."

ME: "That's the bilateral component."

HIM: "But as chief operator I also have to be able to address messages to the passengers, even when they aren't asking for anything and are just sitting there without a care in the world."

ME: "No problem—that's the directional component. I can certainly let you have that," I said in a conciliatory tone.

3.2.2. Track Equipment

This consists essentially of a transmission module of the type used in the metro systems in Paris and Lille, placed in the axis of the track and including the various transmission loops necessary for the ground-to-car connections:

—a continuous-broadcast loop serving as support for continuous ground-to-car communication and programmed intersections that serve to localize cars in space;

—a continuous reception loop serving as support for continuous car-to-ground information.

HIM: "At least they aren't demanding independence, autonomy, consciousness, autocontrol, or pronoun forms. They transmit, period," Norbert noted with satisfaction.

3.2.2. Fixed Equipment . . .

3.2.3.1.1. The UGT (traffic-control unit):

The track is divided into transmission sectors with a maximum length of one kilometer. Each sector is controlled by a UGT that has five main functions:

—it controls the anticollision function between pairs of cars in the sector and between adjacent sectors;

—it ensures the retransmission of telemeasurements and remote-control instructions exchanged between the PCC and the cars located in the sector, as well as transmitting phonic communications;

—it regulates the separations, mergers, and linkups of the pairs of cars;

—it maintains surveillance of the cars present on the sector, in order to supply the PCC with information it can use immediately to carry out its "vehicle tracking";

—it supplies alarm signals to the PCC plate, in order to cut off the high-tension power supply.

ME: "You have to delegate them something else, otherwise they'll drown you in information."

HIM: "No, no, that's out of the question. I'm only delegating the designation of the head of the line to them. I'm taking everything else back."

ME: "You'll never make it. There'll be too much information."

3.2.4 The Central Command Post

The PCC includes the various elements necessary for the supervision of the Aramis system and of the passengers using the system.

It consists of:

—a computer system . . .;

—workstations that bring together the means available to the PCC operators;

—consoles supplied with functional keyboards and color video screens that allow the operators to survey and modify the technological status of the system's components;

—consoles allowing video surveillance of the stations and ensuring the various phonic liaisons with the passengers . . .;

—control panels, if visualization on the consoles turns out to be insufficient; these provide operators with a synthetic view of the system state (position of the trains in the network) and of the state of the energy-distributing equipment;

—telephone stations putting the operators in contact with the various operational poles (workshops, stations), and, through direct lines, with police headquarters, fire stations, and the emergency telephone system.

HIM: "Well, here at least I'm on my own turf. A panopticon, buttons, calls, alarms, control panels, a command post. It consoles me for having given you so many rights. Now it's a real general-headquarters command post. It's not going to be a question of playing the smart guy."

ME: "You don't realize it, but you're going to break down every four seconds. Don't forget that you have 660 pairs of cars to manage. And at the slightest warning I go into security mode and block everything. You were talking about crowds. Imagine what you'll get if we add up all the relationships among all these beings."

HIM: "It's not possible," Norbert exclaimed, terrified by a complication that delighted me as it had probably delighted the Matra engineers. "You can't manage crowds like that. You're sure that this is the price of autonomous existence?"

"Of course; it suffices to make the mobile units more and more intelligent. You'll see—they'll end up managing all those multitudes."

[INTERVIEW EXCERPTS]

M. Laffite, RATP engineer in charge of automation:

"I always forget to mention it, because it strikes me as self-evident, but each car possesses *a representation of the entire track.* It has the invariants—its own number first of all; it knows *who it is,* and this is cabled; it's been given the maximum speed limits for each section; it knows the profiles of all the stations.

"I forget to say it, but this is the base. This is how *it knows* where it is.

"Okay, these are the invariants; this stuff is entered into memory once and for all.

"So, next *it reads* the information module that's on the track. There are two

Figure 14.

modules, reference points every 2.20 meters, perfectly regular so as to keep checking the phonic wheel. Then there's the supplementary module, consisting of irregular intersections that convey information.

"It's like reading a bar code?"

"You might call it that, yes; the track is a bar code *[see Figure 14].*

"The vehicle *learns by itself,* from the track, to locate itself. Nothing is done by remote control. Matra never wanted to use remote control—for example, to have the localization sent by the UGT or the PCC. Remote control is much too dangerous. The car has to know where it is *on its own.*

"So it *knows* its own identity, it knows its position, and from this *it deduces* its own speed and acceleration. At intersections the situation is more complicated, because the platform is interrupted. So there are a lot of little problems to be resolved.

"If a car is lost, it's no longer reliable and here *it can become dangerous.* Sure, obviously, the car can tolerate a mistake, because if the constraints are too radical, you stop all the time and nothing works.

"It transmits this flood of information to the UGT that oversees all the cars in its sector. The UGT sends back all the information to all the cars, and each car

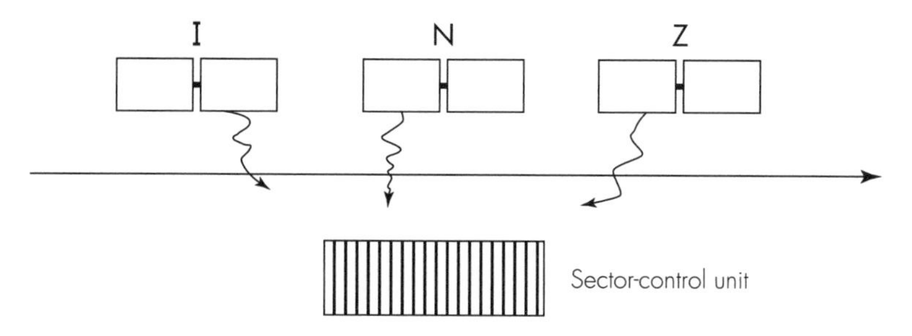

Figure 15.

retains only what concerns itself, and what informs it about its predecessor. That's all; it lets the rest go by.

"There is a security measure here, a major one. Each pair of cars says, 'I have to be interrogated every two seconds. If at the end of three seconds the UGT has not interrogated me, *I stop.*' So very quickly you reach the limits of the UGT's capacity, which has a very limited output in any event and which nevertheless extends over a kilometer with potentially dozens of pairs of cars. So the pairs have to *work things out* on their own *[Figure 15].*

"Vehicle N knows that Z is ahead because the UGT *has told it.* This is the main job of the UGT—to tell N that Z is its *target* and that N has to hook up with it. Okay, N also knows that I is behind, but it doesn't calculate that; there is an ordered relationship, it's up to I to work it out in relation to N as N does in relation to Z. It calculates and works things out. Next N itself makes the calculation about the linkup with its target."

"But why not also delegate to the vehicle the job of choosing its own target? Wouldn't that be more practical?"

"No, that's impossible because of the track. Think about it. There are branchings. How can one vehicle know whether the one on the other track is ahead of it or behind? It knows the abscissa, the *x*'s, but not the ordinate, the *y*'s.

"The viewpoint of the ground, of the observer on the ground—of the UGT, if you like—is the only one from which the notion of target can be decided. So this is the idea: all the information is sent through ground-to-car transmission, but the only relevant information is *the assignment of a role:* 'Hey you, you're the target; not you.'

"And then the cars themselves send each other all the necessary information through the direct car-to-car hyperfrequency link. This one is refreshed much

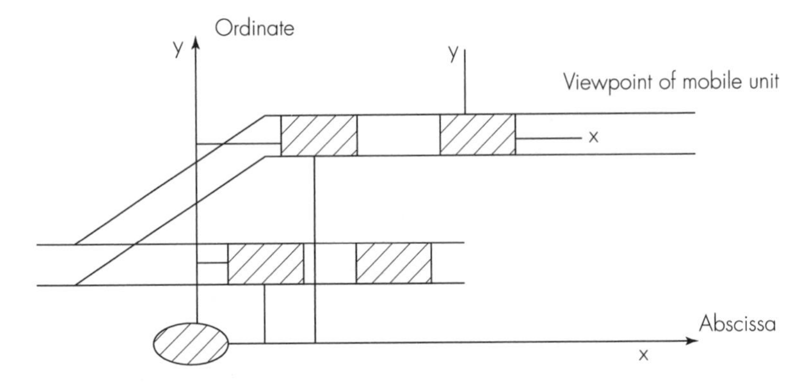

Figure 16.

more often—every 20 milliseconds—so it allows much greater finesse. But as for whether a given car is to be a target or not, it's the UGT that *assigns the role,* that makes the decision *[Figure 16].*

"Doing this already entails quite a few problems, because the time needed for real-time calculations with all these accelerations is not easy to manage, and in addition everything has to be coded, *encrypted,* to avoid errors; it's as if everything were being done twice, once openly and once in cryptic form, and then you had to compare the two.

"If they match, you say, 'Okay, the order can be carried out.' If they don't match, you send an alarm; you suspend the action and you check the calculations.

"In addition, there's a timer that ensures that the information is updated, to make absolutely sure that the information is current and *hasn't been lying around for several milliseconds.* But all that takes time, space, personnel.

"Matra decided to do all that in 16 bits* to gain speed. At first they started off with 48 bits, but it was too slow. But with 16 bits, they can't possibly have security." [no. 16, p. 9]

---

*The length of elementary messages. The shorter the message, the quicker the calculation; but since, to gain security, the information has to be encrypted in parallel form, half the bits are needed. If the message is longer, in order to allow security, the whole set of calculations is slower.

"You think we're going to have to go into the bits?" Norbert asked me, sounding more and more worried.

"Follow the actors, my dear man, follow the actors. Those are your methods, right?"

"But the farther we go, the more crowded it is. Every part of the system is as complicated as the system as a whole. Every plate we unfold is itself made up of plates to be unfolded!"

"Pure Borges, my dear mentor. Why let it upset you? You love literature, after all—and you love folds."

"I'm not upset," he said stiffly. "I simply have the feeling we're getting bogged down."

"The only question is whether the details are strategic or not. And these particular ones are important; they're at the heart of Aramis' autonomy, the adjustable mobile unit. If it can do that, it's an autonomous being, a real automobilist. It can exist."

"A heteromobilist, still; we're the ones who give it its laws."

"No, it becomes autonomous for real; we've given it its laws for all time."

"But it's like us," Norbert snapped back furiously. "If we were characters in novels, we wouldn't escape our author."

"We would, too, just like Aramis. That's what we used to say when we were kids: 'You gave it away, you can't have it back.' Look, you taught me yourself, the creature certainly escapes Victor's control, and the Frankenstein character escapes Shelley to the point where the monster that was created is called by the name of the monster who created it."

"But that's just the point," said Norbert irritated by my new mastery of the project. "We know now that Shelley wrote nothing but lies. None of them can escape on their own. They need a following, company, a crowd. Even Adam can't go on alone. Even to sin, he still needs grace."

Theology allowed Norbert to hide the fact that he had lost his grip and could no longer write any "sociological commentary," as he pompously called it, on what we were discovering.

The next plate administered the coup de grâce *[Figure 17]*.

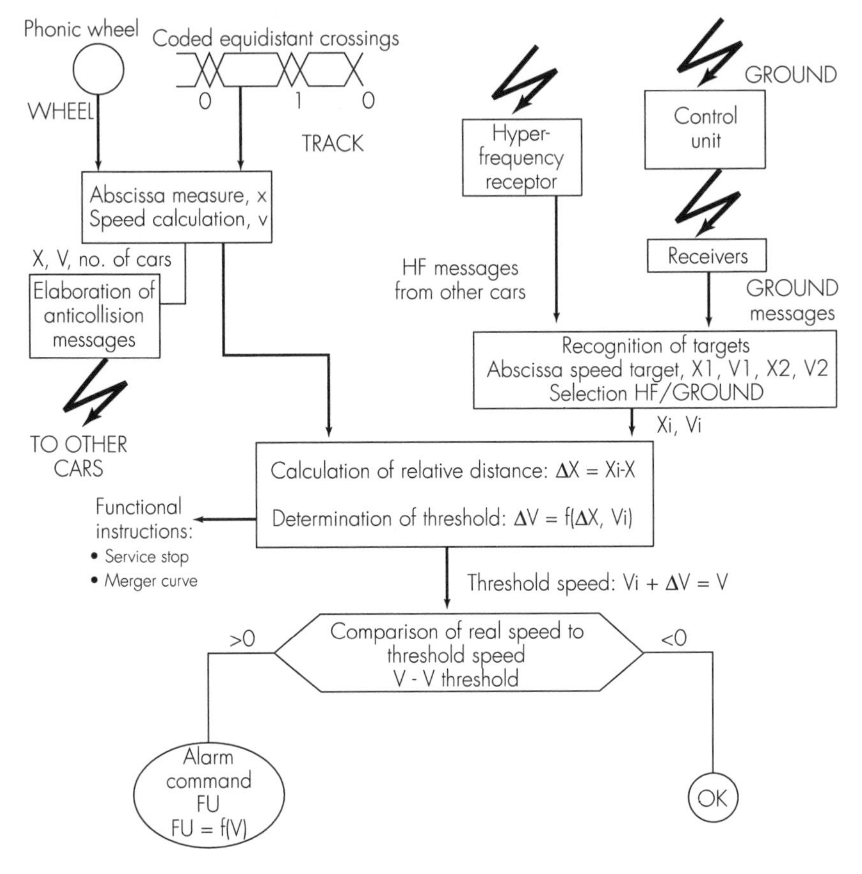

Figure 17. Anticollision safety-control principle (Document LB66/AR/E400 18/86/SA/NT, Figure 4.4.2).

The anticollision system is based on the principle of the CMD (adjustable mobile zone), which means that the free zone (the minimum security zone separating two sets of successive vehicles) is *mobile* to the extent that it follows closely the displacements of the vehicles (as opposed to detection by fixed sector) and *adjustable* to the extent that its length varies as a function of the vehicles' speed.

"It's just like in an automobile," I had to explain patiently to Norbert. "You can follow the car ahead more closely if you're going more slowly; when you're stopped, you can almost touch it. With Aramis, the advantage is to be able to bump a bit, even in motion, thanks to the shock absorbers."

This anticollision principle leads to ensuring, at the level of the vehicles, the following operations:

—absolute and instantaneous measure of the abscissa and the speed;

"It's like reading the speedometer in a car, 20 kilometers/hour, 40 kilometers/hour."

—communication of the coordinates (abscissa/speed) of the target vehicle;

—calculation of the relative distance separating the vehicle from its target;

—elaboration of a security speed threshold not to be exceeded, a function of the relative distance and the absolute speed of the target;

"If you're 100 meters away and going 50 kilometers/hour, it's more dangerous than if you're 100 meters away and going 10 kilometers/hour."

—irreversible emergency braking command if the speed threshold is exceeded.

"Here deceleration has to be factored in. The faster you go and the closer you are, the faster you have to slam on the brakes, but you have to make sure the guy behind you knows you're slamming on the brakes. You have to warn him—it's like the red brake-lights.

To achieve this,

—every vehicle always knows the number of the pair of cars ahead (target vehicle);

—every pair of cars measures its own abscissa in relation to the origin of the sector by counting the cogs of the phonic security wheel mounted on every right front wheel. This measure is periodically recalibrated through detection of the recalibration intersections, which are evenly spaced every 2.2 meters;

—every pair of cars calculates its own speed on the basis of the count of the number of cogs in the phonic security wheel;

—every pair of cars transmits to the ground (UGT) and to the rear (through the average distance car-to-car hyperfrequency liaison) its own position, its speed, its number, and its alarms, if any, in the form of a coded message;

—the UGT rebroadcasts the information received to all the pair of cars;

—each pair of cars calculates the relative distance (Δx) separating it from the preceding pair through the difference between its own abscissa and the abscissa of the target (transmitted by the UGT or the hyperfrequency liaison);

—each pair of cars calculates a threshold speed corresponding to a safety zone in case of emergency braking;

—each pair of cars compares its own real measured speed to the threshold speed. If the former is higher, the pair orders an irreversible emergency braking action." [p. 35]

"And if I give you all that," asked Norbert feverishly, "you're autonomous?"

"In any case, I am an automobile without a human driver. Obviously I'm not all alone. I have the rails, six onboard computers, three antennae, an encoding service, a decoding service, ten dozen inspectors."

"But then you're moving? You're alive?"

"Let's say that I'm opening one eye. I'm stammering. At the slightest alert, I shut down everything and become inert again. I fall into security mode."

"And the other one, Shelley, with her monster that goes off on its own to visit Germany and Switzerland without even unhooking its intravenous tubes! And that learns to read English by studying Milton's *Paradise Lost* through a crack in the floorboards!"

"Yes, I didn't want to say anything, Norbert, but that was a little hard to swallow."

"Are you still as sure as ever that the CET engineers are going to do away with all those multitudes? If we declare this phase innocent as well—and this time, you're the one who'll do it—we have nothing left to turn to."

"I hope so. In any case, they've done a good job. They're the ones who've written *Paradise Lost*. And not in verse."

"In what?"

"In program lines."

"We're going to lose our hides here," said Norbert, alarmed by the unanticipated turn I had given to the investigation. "You work on the technological specifications. I have to write the report."

```
                    code 5800.2
  ORI |0 0 0 | 00 111 100                   |0030FFFF      #8
CCR
  ORI |0 0 0 | 01111 100   |0070FFFF   #16 SR
  ORI |0 0 0 | bwl ae ae   |0000FF00   #bwl ae4
AND1  |0 1 0 | 00 111 100              |0230FFFF  #8 CCR
AND1  |0 1 0 | 01 111 100              |0270FFFF  #16 SR
SUB1  |0 2 0 | bwl ae ae   |0400FF00   #bwl ae4

                      bwl
                      00         |B
                      01         |W
                      10         |L
                      00         |llgl
```

I didn't see Norbert for the next two weeks.

Norbert told me that he'd dreamed he'd found Victor Frankenstein on an Alpine glacier, just when the Creature had finished its horrible story. And he had interceded on its behalf!

N.H.: "'You're deceiving us, Victor, and you've been deceiving us for a long time. You bewail your crimes, but you do this in order to hide another, bigger crime. Your sin is not that you have fashioned the monster. You created it for its beauty, for its greatness, and you were right. Your crime does not come from the hubris of which you accuse yourself; it is not that you played the demiurge, that you wanted to repeat Prometheus' exploit. Your crime is that you abandoned your creation. Were you not the first person it saw when it opened its eyes? Did it not stammer out your name? Did it not hold out those deformed limbs toward you? It was born good, like you, handsome like you, wise like you, since you were its creator. Why flee? Why leave it alone, ill adapted to a world that rejected it? That is when it became wicked! And even now, you turn your eyes away from it. You are horrified by it, even though it is yourself twice over, you in your beauty and you in your cowardice, you in your marvelous creation, and you again in your shameful flight. No, your sin is not that you took yourself to be God, for God never abandons his creatures, no matter how sinful. He follows them, sacrifices Himself for them, throws Himself at their feet, sends them His only son; He saves them. Continual, continuous creation. Salvific incarnation. Your inexpiable sin is not that you continued to play God. You shouldn't have begun in such pride, only to finish with such pusillanimity. You shouldn't have begun by loving, only to end up hating so much. Look at the poor thing—horrible, yes, of course it is. But what is sinful Adam in the eyes of his Creator? A much worse monster. If you have to expiate something, at least have the courage of your own crime; don't hide behind the pitiful excuses of the sorcerer's apprentice. Your creature didn't escape you, for you and you alone chased it out of your laboratory. Oh, the accursed one who mistakes the curse! The forked one who turns away hatred and who continues to deceive at the very heart of his repentance! Through your fault, technologies lie accused, abandoned as they are by their creators, by all the Victor Frankensteins who take themselves to be God on Monday and ignore their creations the rest of the week. And he preaches modesty! 'May his pitiful history,' he moaned, 'at least serve as a lesson to all would-be demiurges!' But you're drawing the wrong lesson. It is not our creative power that we need to curtail; it is our love that we need to extend, even to our lesser brothers who did not ask us for life. We acquainted them with existence. We need to acquaint them with love. And what else is it asking you for, the monster that is imploring you to make him a companion in his own image, if not for love?'

"Victor, who had been shielding himself from the sight of the creature, lowered his hands and looked at me. Then, slowly, he turned and looked at the monster lying in the snow. The glacial cold was suspended. The two beings, their eyes full of tears, were now looking at me.

"'Yes,' I said, 'it must be done.'

"Victor made a halting gesture of reconciliation and approached the monster who had lowered its head; he stretched out his hands and his lips as for a kiss. At that moment, he drew back in horror and, screaming, fled once again.

"I remained alone on the sea of Ice with the creature. It was my turn to approach it. It offered me its horrible, pitiful countenance."

Norbert woke up terrified.

The next morning, he arranged to take sick leave. As for me, left to my own devices, I plunged into the CET codes, swimming along after the Matra programmers and taking as much pleasure in the task as they had.

7 ARAMIS IS READY TO GO (AWAY)

When I went into the laboratory director's lavishly appointed office, I knew right away that something was wrong. Norbert H. was not available, I was told. But given the reputation of the team and the rapidly approaching deadline, even though my status as a simple intern made the request unusual, still, the recent discoveries of my professor, and with all the interview notes it would be easy, and Norbert thought very highly of me . . . In short, I was going to have to draft the report myself, in a week's time.

"And since you're an engineer," the director told me, "I'd like you to make a special point of finding out whether or not it's technologically feasible."

I had to dig back through the files on my own, and go back over the transcripts of all fifty interviews. I found out that my professor's archives were not nearly as well organized as the RATP's.

[DOCUMENT: RATP, "RAPPORT GÉNÉRAL SUR LE DÉROULEMENT DE L'OPERATION," JANUARY 1988, P. 53: CONCLUSION OF THE LAST OFFICIAL DOCUMENT WRITTEN AT THE END OF PHASE 4, AFTER THE ARAMIS PROJECT WAS TERMINATED]

The results of the CET Aramis-boulevard Victor operation are the following:

—The team has mastered microprocessor-based automation techniques as applied to automated transportation systems.

This is the case for the Matra company, which is currently using a major portion of the Aramis team for the operation of Line D of the Lyon metro.*

—The vehicles, in particular the P3, P4, and P5 prototypes, are nearly ready to go into production. On the whole, they can be termed a *success,* pending a few still-needed improvements.

—The use of Aramis' variable-reluctance motor, approved by the RATP, represents an *important innovation* as a mode of propulsion.

—The studies and the production of the adjustable mobile sector carried out during the Aramis operation will be followed by the refinement of comparable technologies for the Lyon metro.

—The studies of ultrasonic and hyperfrequency transmissions may give rise to *future developments* of interest to the world of transportation.

In its separate pieces, Aramis is feasible and the engineers are happy with it, I told myself. The motor, the car, the mobile sector, the connections—all these work well and can be used again elsewhere. Thus, Phase 4 is positive: Aramis is viable, and the deal† has been upheld. Only one thing left to do—namely, bring together all these viable elements and get Aramis into production by creating the line of which the CET on the boulevard Victor is the head. Norbert was on the wrong track from the very start. Aramis is a fully developed technological invention abandoned by the politicians just when everything was ready for the manufacturing stage. It's a story of the Concorde variety—technologically perfect, but in need of a massive political push to survive.

*The traditional subway in the city of Lyon has recently been supplemented by a new line that is much like the automated VAL.

†Still in the contractual sense (the French term used here, *marché,* can mean "market," "deal," and "contract").

IV.2.2.1. Functional Observations*

IV.2.2.1.1. General Framework

The verifications of the functional and technological status of the equipment at the time the experiment was terminated took place from Monday, November 23, 1987, to Tuesday, December 1, 1987, inclusive (8:30 A.M.–5:00 P.M. weekdays). Of these seven consecutive working days, two were neutralized so as to allow Matra Transport to get the P1 vehicle to function as anticipated; despite this additional delay and the initial postponement of the starting date for the trials from November 2 to November 23, the P1 pair of cars was unable to participate in the trials . . .

The verifications to be carried out by the representatives of the RATP were defined by Matra Transport, which prepared documents for this purpose describing the anticipated trials within the framework of the codicil. There are fifteen of these "statements." Each one in fact corresponds to a reduction of the PVRI—the minutes of acceptance trials for proving compliance with the specifications set *[cahier d'essais de procès-verbaux de recette individuelle]*. The PVRI, having become statements by reduction, were thus insufficient to "qualify" the functions or the subsystem . . .

IV.2.2.1.3. Results

The verification log intended to record the conditions of the trial is a transcript of notes taken on that occasion. It describes in chronological order the unanticipated incidents that occurred during the verifications . . .

Without claiming to deduce from these figures anything other than trends, let us note that the average time between incidents was approximately ten minutes.

*In the framework of a contract, the contracting authority *[maître d'ouvrage]* is required to provide legal verification that the commitments made by the contractor *[maître d'oeuvre]* have been carried out. Receipt and homologation are two steps in the process of legislating the technological capabilities of a transportation system. Verification is accomplished by contradictory procedures that have to be approved by both parties.

This figure, in the context of the initial objectives of the contract (300 hours for the complete equipment), indicates the efforts that would have still been required to reach those objectives.

IV.2.2.2. Endurance Trials

Adjustable mobile sector, three pairs of cars. Number of hours anticipated: 200. Number of hours in CMD: 15 minutes. Number of incidents: 3.

Function of linkup, merger, demerger: with two pairs of cars. Anticipated length: 4 hours. Duration of the trial: 3 hours 15 minutes. Number of hours during which pairs of cars were in linkup, merger, or demerger status: 43 minutes. Number of incidents: 29. Number of successful linkups compared to number of attempts: 6/13.

The very high number of incidents shows that the system was far from perfected . . . [p. 46]

1. The cost of the pairs of cars is roughly 30 percent higher than the target figure, whereas there was an allowance of 20 percent.

2. The cost of onboard automation was significantly underestimated at first (+70 percent); it now represents more than 28 percent of the cost of the rolling stock.

3. The cost of fixed automation is in line with the target figures. [p. 52]

I don't get it. They've been telling us all this time that Aramis is feasible, yet three pairs of cars can't hold the road together for more than fifteen minutes, and even so they stop three times, and the security systems aren't in place! And it was going to take 600 of them to outfit the Petite Ceinture! It's the story of the guy who wants to go to the moon, who climbs up a stepladder and is sure he's on the right track, because "all that's left to do is extrapolate." When there's enough food for three, there's still not enough for 600, barring a loaves-and-fishes miracle. But if the extrapolation from two to 600 is impossible, then it's not like the Concorde. And there are the skyrocketing prices! Still, the conclusion of the contradictory report is positive. Ah, now I understand why it's called a "contradictory report" . . .

[INTERVIEW EXCERPTS]

M. Trouvé, an important technological director at the RATP. "It's easy enough to show that it's feasible with three cars, but it's a completely different matter to say whether or not it can be produced and implemented. We started with a good automation system, with the two-pair train; and it more or less worked with three.

"We told ourselves, 'It's almost there,' yes, but when we got rolling, it was without safety measures, and when we added those we were in the manual mode, not automatic . . . It's not the same thing.

"The first phase may be misleading. Matra relied on it a lot, moreover—'It was almost ready.' I had to fight to keep the codicil from declaring that it was homologated.

"In fact, the codicil was a disaster. Matra wanted to turn over the key; we insisted on winding things up properly. I have to say that Matra also made it a point of honor to go further; they stopped on December 11 instead of November 30.

"But finally, they didn't bring off the demonstration. I personally insisted on seeing three pairs of cars running together. They would have yelled, 'Hooray for us!' They did their best; they didn't make it.

"But *with two pairs, they achieved a good level of reliability.* They transported Chirac and Balladur, without stressing the safety issue."

"But in your judgment, was it technologically viable?"

"There was the complete technical review of 1982–83; we had concluded that it was viable. I continue to think that it is, but in a time frame that I have trouble extrapolating—and anyway, it's a luxury system, it's the Concorde of public transportation . . .

"Let's say *there were no factors precluding* the existence of Aramis; that was the opinion of the technology committee. We couldn't see the end of the tunnel, but *we had no reason to think that it wasn't there.*

"The problem is that even if they'd gotten an extension, Matra's directors wouldn't have gone on; they'd lost interest in the problem. Builders have a short-term vision. 'As long as there are no plans for the line, I'm not going to invest in Aramis'—that's what they were telling themselves." [no. 5]

He continues to think that it's feasible, even if he's less than enthusiastic, so maybe I was right. But even if Aramis has been perfected technologically, it's too expensive, and in any event the manufacturer in charge of producing it didn't give a damn. A peculiar Concorde. What does it mean exactly, something that's doable and that isn't done, that nobody wants to do, that nobody can do, that nobody knows how to do, that nobody can afford?

[INTERVIEW EXCERPTS]

M. Alexandre, Matra's head of the CET, speaking at Matra headquarters:

"From July to November, we'd made so much progress, we would have gotten there; I don't know how long it would have taken, or what we would have needed. The software was extremely complex, but at the hardware level we didn't have many reliability problems. We would have carried out the contract in the long run."

"So, in your heart of hearts. . .?"

"We would have gotten there, but it would have cost more than two million francs."

"So it's a little like the Concorde, hyper-state-of-the-art, feasible but not cost-effective?"

"Yes and no. We could have pulled it off if we'd been more determined. We should have looked for outlets that were less solvent but more political. We should have started in Montpellier, for example; we should have experimented commercially and moved *later on* to using multicar trains and link-ups. We should have taken a more gradual approach.

"But the CET goals were too rigid.

"The mechanism for maintenance support was a luxury.* Of course the software could do the diagnosis, but it's another matter to go from there to

*Autodiagnosis would have led to a reduction of the enormous maintenance tasks involved in Aramis, once the system was built.

having each car say, 'Here's the problem, here's where I'm in trouble.' We didn't say so to the RATP, but we thought that was useless." [p. 11]

"You would have gone further if you'd been able to renegotiate the contract?"

"You have to see that there was this organizational abscess, I mean this organizational aspect. We could have held onto each function without degrading it, but by taking the bottom of the line every time to absorb, for example, Montpellier's 50 cars and only later moving to the Petite Ceinture's 600. For example, if we'd been able to move to 200-millisecond updating and leave it at that, we could have had less sophisticated software, and that would have allowed us to do Aramis; we'd have been downgrading, relaxing the constraints.

"Well, I could never get Parlat *[his opposite number at the RATP]* to agree to that. 'We'll refer to the contract when you see us really go off the deep end; you're holding up a guard-rail, but while things are going along normally you forget that and we manage things with a certain amount of flexibility.'

"But the RATP guys had the impression that we were constantly backsliding in relation to the objectives." [no. 25]

Matra's M. Frèque:

"Let me show you something that will help you understand" *[see Figure 18]*.

"Of course it's a bit exaggerated, but actually things often go like that. Why do something simple when you can do something complicated?

"The RATP was always asking for something slightly extravagant. You start with some simple doodad, easy as pie, and you end up overspecifying; you get something that says 'Mommy, Daddy,' but it obviously costs more.

"Besides, yeah, this is really classic, there was always somebody who'd say, 'We did that in Lille'; it was team culture. Habits count for something as well." [no. 6]

Yet another interpretation! It was almost perfected technologically, or at least it all could have been if the RATP hadn't insisted too rigidly on impossible constraints. Yet then what they would have perfected wouldn't have been Aramis, but something else, an Ara-*x*! The operating agency says it's feasible but claims the manufacturer has abandoned the baby. The manufacturer says a different Aramis was

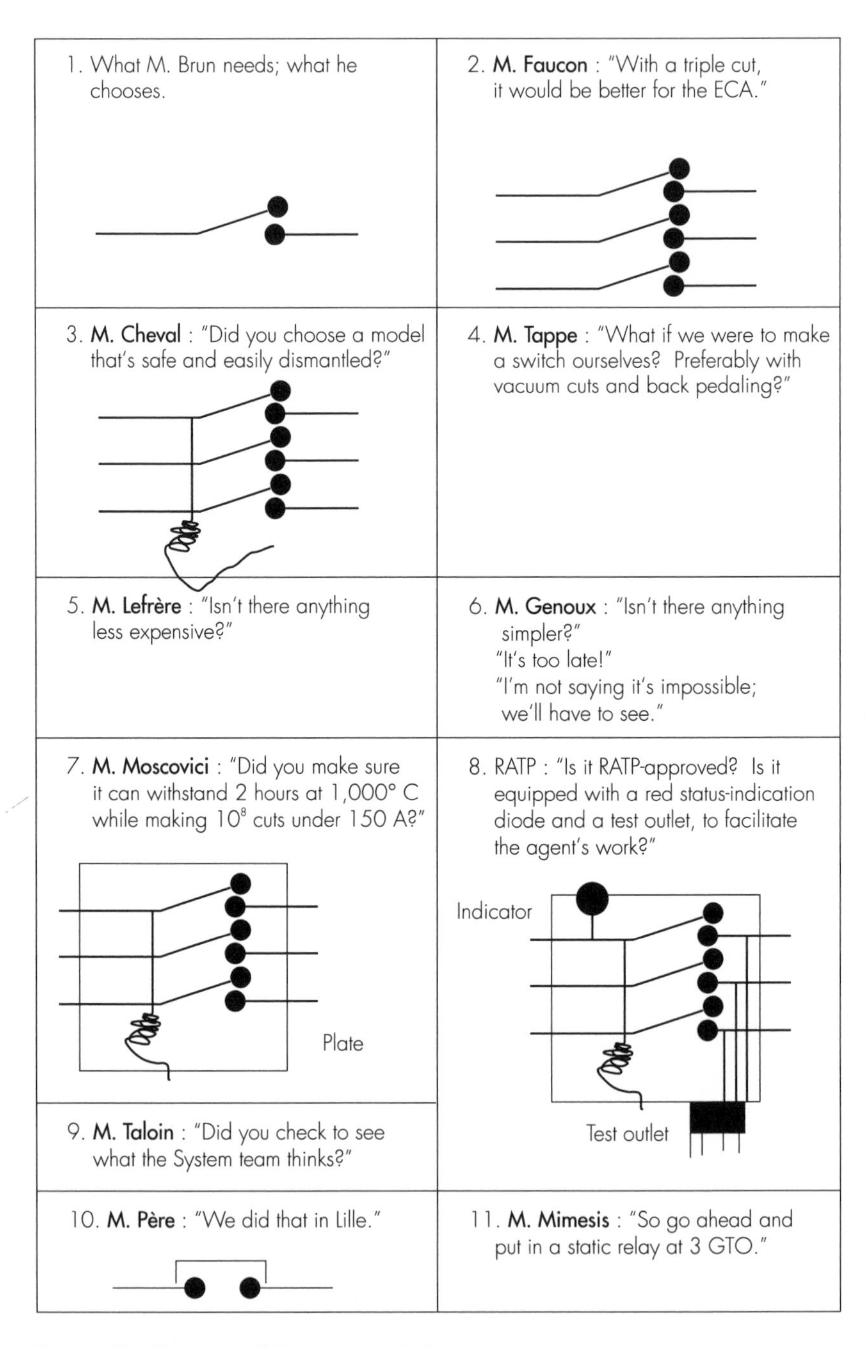

Figure 18. The story of the Aramis switch.

feasible but claims the operating agency was asking the impossible. I'm really starting to wonder whether Norbert wasn't right. It's not so easy to decide whether Aramis is technologically feasible or not.

[INTERVIEW EXCERPTS]

Messrs. Gueguen and Parlat, of the RATP, who participated directly in the project and in the drafting of the end-of-contract contradictory report:

M. Parlat:

"The technicians weren't able to stand up to the politicians. The current line is that 'technologically it's a success'—but that's not the way we see it."

"So the CET wasn't useful?"

"Oh yes, the CET had to be done, it was indispensable; with studies, on paper, you never go far enough. You say, 'We'll get there,' you think problems always get solved, you think engineers can always muddle through, but with the CET we really had our backs to the wall."

"You know that you're being accused of rigidity, of being unwilling to simplify Aramis, of insisting on respecting the contract too rigorously and of having scuttled the project for that reason?"

M. Gueguen:

"But that's not being rigid—we didn't want a mini-VAL! If you give up nonmaterial couplings, it's not Aramis any longer. There weren't any development problems with the mini-VAL; it exists. Among ourselves we were calling it an ARAMIS-VS, for 'very simplified!'"

"Yes, but if it wasn't technologically feasible, of course you had to simplify."

"It's more complicated than that. The people in charge should have said 'stop' if it was infeasible; *we* knew there were problems, but they shouldn't have come around telling us it was technologically perfect, when Matra hadn't perfected it.

"In *Entre les lignes* I read something interesting *[see the second document in the Prologue]*: 'It's a technological success, but there's no application for it and it's not our fault. If we kept going we'd be ready.' How do you expect us to believe that?

"Obviously, *we can be accused of not having tried to call attention to the problem;* but we did try, and our concerns weren't taken into account . . .

"You have to understand why they all say, 'It's a technological success,' because if someone says, 'After fifteen years, we have nothing,' how can you expect them to accept it? They're going to wonder, 'Where did the money go?'

"So they say that all is for the best in the best of all possible worlds.

"Over here, we say what we think; we'll see what management does. If a politician says, 'It's okay, it's all going very well,' you know it's cheating, but to say that they stopped because there wasn't a line is a form of cheating. I don't understand how technicians can go along with that. Obviously there's also the cost. They say, 'It's too expensive,' 'It works, but it costs too much,' but nevertheless it's technologically state-of-the-art. *It doesn't work as well as that.*

"You can't even say that it's for later, for the future, because if Aramis had to be started up again, now, *everything would have to be done over.*

"No, the adjustable train—we don't know how to do that, short of a technological revolution. And to meet the specifications, we have no idea what it would have cost, or how long it would have taken, just to get to five pairs of cars, and from there to 600 . . .

"At the same time, it's too bad to have gotten so far and then say, 'We're shutting it all down.' But if it's so unreliable, why had it gone so far? And if it's perfected, why not go on? I personally wouldn't like to be in Alexandre's place *[his opposite number at Matra].* He was saved by the bell. He must have seen the technological difficulties the same way we did. He must be annoyed that it ground to a halt the way it did, but at the same time he must be happy that he didn't have to meet the specs—yes, he must be really relieved.

"For our part, we did the contradictory report; we told the truth. We weren't about to go ask Matra, 'And what do we put here?' That's for management to worry about." [no. 10]

Saved by the bell! That's not what poor Aramis would say. Finished off, on the contrary, by a fratricidal struggle between the two teams, the contracting authority and the contractor, with one accusing the other of downgrading Aramis' principles, and the other retorting that it was being required to pursue impossible objectives whereas everything would be viable if the constraints were relaxed. And the struggle doesn't even show whether the thing is technologically feasible! It's an insoluble problem. I have a feeling I'm going to be in hot water with Norbert's director.

[DOCUMENT: LETTER ADDRESSED TO THE TRANSPORTATION MINISTRY BY M. DESCLÉES, PRESIDENT OF THE DEVELOPMENT COMMITTEE—THE BODY RESPONSIBLE FOR KEEPING SYSTEMATIC TRACK OF ARAMIS ON BEHALF OF THE MINISTRY]

Dear Director,

You asked me to find out what the Aramis development committee thinks about the modification of the experimental program.

The committee met at my request on September 30, and I am sending you a report that conveys the views of its members.

Let me add that the program is being terminated at a point when there seems to be *no serious obstacle* to perfecting the system, yet at a point when not all the innovations can be fully demonstrated. This is surely regrettable. Nevertheless, *the personal conviction* of the committee members is that this system could be brought to a level of technological realization close to what was anticipated in the protocols and in the RATP contract. However, the committee raised further questions about the conditions for operating a complex system of the PRT type; but only commercial trials would permit a definitive judgment . . .

The majority of the committee members *remain convinced* that a market for Aramis exists, that halting its development is not justified from this standpoint, and that it would be opportune to undertake an *a posteriori* evaluation that would make it possible, in particular, to draw up a contract and to see whether it would be appropriate to modify certain functional specifications with respect to those of the contract.

Very truly yours,
Desclées, Director of INRETS

Ah, but this changes everything. The engineers on the development committee who studied Aramis from the beginning agree: Aramis as a mobile unit has been perfected technologically, the contract was nearly fulfilled, only the operational side is left with a few unsolved problems.

Well then, that's reassuring. I started worrying too soon. The engineers closest to the project have doubts about its technological feasibility, for psychological reasons perhaps, but the experts who are farthest removed remain quite satisfied.

[INTERVIEW EXCERPTS]

M. Desclées, author of the foregoing letter:

"My only worry was that the political decision *[to quit]* should not damage what had been gained scientifically, and that they should say, 'It's not because it didn't work that we're stopping.' That was my concern . . .

"I didn't want Matra to be held responsible; I didn't want it to be said that they had to stop because Matra hadn't met its obligations, you understand?

"I didn't want to cast any further discredit on the technology, anything that could have given the Budget Office more grist for its mill *[to criticize other innovations in public transportation]*."

"So it's out of solidarity for those who were defending technological innovations? It's tied to the fact that Fiterman leaned rather heavily on the Budget Office?"

"Things like that are never forgotten; hence the orderly retreat." [no. 11]

But this changes everything once more! The Budget Office again! How am I expected to decide whether Aramis is technologically feasible or not when they're arranging, consciously and deliberately, to protect Matra in order to preserve the future of huge, costly, technological projects? The question of technological feasibility is muddled for muddling's sake. What? Could there be a conspiracy to protect the engineers? Are they behaving like schoolboys who've done something foolish and don't want the Budget Office to catch them?

Ouch! I have a feeling I'm getting confused again—starting to do bad sociology, to use a borrowed phrase. I'm going back to blaming and denouncing. It's also because Norbert is letting me work all by myself. Instead of running after monsters to kiss, he ought to be doing the technology audit himself . . .

I'll have to see what the press file says. With all those journalists who swarm around juicy subjects, there must be some interesting revelations about Aramis' feasibility.

[DOCUMENT: ARTICLE BY ALAIN FAUJAS, *LE MONDE,* OCTOBER 26, 1987]

By November, the definitive results will be in, but they will not keep the project from being abandoned. Responding to pressure from Matra, M. Jacques Chirac, mayor of Paris, had proposed to install it between the Gare de Lyon and the Gare d'Austerlitz. But on this minor branch line, Aramis would not have shown its full capabilities. The regional officials seem to prefer the classic metro inside Paris proper. As for the RATP, it considers Aramis a technological success and a commercial failure. In its view, Aramis has too little carrying capacity, is too sophisticated, and costs too much for its intended use—that is, transporting a maximum number of passengers. At a time when the RATP is tightening its belt, it prefers to reserve its investments for things like progressive automation of its existing lines. Matra has chosen not to comment on the probable abandonment of the program. The CEO of Matra Transport, a subsidiary of Matra, says only, with a touch of bitterness, "We couldn't do it all by ourselves."

Yet Aramis will not have been tested in vain. It has led to better mastery of automation and improvement of the wholly electronic guidance system that will go into operation on the newest line of the Lyon metro and that will pilot the second-generation VAL, Matra's other metro, the one that has been a success in Lille, Chicago, Jacksonville, and Toulouse and that will win out soon perhaps in Strasbourg and Bordeaux.

Well, it's a good thing there are journalists! The press dossier is unanimous. I've found more than a hundred articles saying the same thing, give or take a nuance or two. Aramis has been perfected technologically, but it isn't profitable; the RATP has changed its mind or has run out of money and is no longer willing, or no longer able, to finance the line; Matra has done a good job; and in any event there are technological benefits for VAL. So here I am at last with a good, stable, new version: Aramis has been perfected, but only in little reusable bits, in separate pieces that are of use to its older brother VAL, always VAL.

[DOCUMENT: MATRA, CONFIDENTIAL INTERNAL MEMO, JULY 2, 1987]

The attached text, developed by the RATP with our approval, constitutes what we have agreed for now to tell journalists if we are questioned:

The refinement and test runs of the Aramis system at the experimental center on the boulevard Victor are going into their terminal phase.

Aramis is making its first runs with electronic couplings on the track; this is the last stage before the multicar train without the traditional mechanical coupling used by trains and classic metros.

The conclusion of these test runs is expected in November 1987.

It will then be up to the four partners—the State, the Ile-de-France Region, the RATP, and Matra Transport—to decide whether or not it is appropriate to go ahead with test runs aimed at full refinement of the system (manufacturing studies and endurance trials).

The prospects for applications of the system are, in the short run, economically less favorable than they were in 1984, when the development contract was signed.

But as of now the development of Aramis has led to increased knowledge and expertise in the realm of automation; this has turned out to be very useful in the refinement of the system of automated guidance that is derived from the SACEM system used on Line A of the RER and that will be applied in Lyon as well as for the second-generation VAL.

But this memo has the very same structure, a few minor variations notwithstanding, as all the newspaper articles I plowed through! What? No independent investigations, no fresh opinions, no research, no criticism? They all say the same thing: what the principal actors have put in their mouths! What doormats, these scientific journalists! Always ready to popularize, never to investigate.

If I ever decide to change careers, I'm going to be a journalist, but I'm going to do investigative journalism, and if I come across a story like Aramis, it won't be my master's voice you'll hear. I'll be the Bob Woodward of technological projects. And to hell with "refined" sociology.

It's obviously a conspiracy! Norbert is completely off base with his idea that Aramis is for once a project without any notable scandal. There certainly is a scandal to denounce! But from another standpoint, he was right after all: there's no way to decide about the technological feasibility—everything is muddled for muddling's sake. But why, oh, why did I get myself into this mess instead of just going ahead to do technology?

[INTERVIEW EXCERPTS]

M. Létoile, engineer at the Lille Research Institute, a branch of INRETS, where VAL was developed:

"Was it feasible? Listen, it all depends. As for the guidance system and linkups that existed nowhere else—there I think the CET pushed the demonstration as far as it could.

"I very much regret that the project wasn't extended for another year. It's too bad they pulled back—in the face of what? Twenty million francs? Thirty million, to do a complete demonstration? At the level they'd reached, they'd proved its feasibility.

"When you went to the site on the boulevard Victor, you'd see the pairs of cars and you really had the impression that they were coupled mechanically. Okay, obviously, there were only two of them, but three, five, it was feasible.

"*I remain profoundly convinced* that they were getting there. But it's true, *the demonstration was lacking;* the appetite was whetted but not satisfied. To me, that's too bad. Obviously, to go from this point to 600 pairs of cars south of Paris—the CET didn't deal with that."

"So for you the CET didn't reveal any serious impediments?"

"No, from the point of view of INRETS, I don't see any insurmountable defects; but then I don't know whether Matra had problems, or if they just ran out of time." [no. 26, p. 5]

No, no, clearly it's not a conspiracy. All the engineers have an unshakable faith in Aramis' feasibility. Obviously they aren't committing themselves as to the costs, or the time frame, or the operational viability . . . These are good examples of real engineers. Their faith is indestructible. Nothing induces them to doubt. Their certainty is perhaps a problem, however. What is a transportation system in which only the mobile unit works, and not the overall operation? And can the mobile unit be said to work if two of them can run together for a few minutes before bumping into each other? That's the motive of the crime, Norbert would say.

[INTERVIEW EXCERPTS]

Two engineers from the Ile-de-France Region, Grinevald and Lévy, responsible for transportation:

Grinevald:

"The ideal for them *[the engineers who developed the project]* is a system with no passengers. The trouble is, there are passengers; you can't avoid thinking about them. What are they going to do?

"And I'm putting myself in the best possible situation, imagining that all the technological problems have been resolved. *[In fact, he thinks this is impossible; see end of Chapter 3.]* But that still isn't enough to make a usable system.

"Obviously you can make a system *called* Aramis that works; what I'm talking about is a system that really brings Aramis to life.

"Lets go into details: even with two or three branches, as you have on the Petite Ceinture, you need a range of missions, plus a high frequency, plus a fine-tuned adjustment of supply and demand. Now all this imposes contradictory constraints."

Lévy:

"It's clockwork; the smallest grain of sand brings everything to a halt."

Grinevald:

"Yes. If we keep Aramis, *it only becomes feasible only if it stops being Aramis.*

"Of course, if it's a big VAL, then it's not a problem.

"Look at the users; they're already lost in the RER,* where there aren't a lot of destinations to choose from. And Aramis doesn't pull up along the entire platform. You add the waiting lines and elderly people—if just 5 percent of the people are lost, you have chaos. And the interval between trains is 45 seconds!

"We were told, 'We're going to deal with that through the experimental method, by trial and error.' And in the end, we're left with a single branching!

"Up to now, I've been talking only about the public, about the interface between the public and the system. Beyond that, there has to be an adjustment of supply to demand. There's a flow matrix to be satisfied that has never before had to be resolved under such complex conditions.

"In addition, Aramis takes only seated passengers, which means giving up all flexibility; there's no buffer for adjusting supply and demand. Besides that, there are families, groups that won't want to be separated! . . . And besides that, the demand has to be regulated constantly, because it evolves, and there are no statistics refined enough to deal with it.

"You have to be able to accommodate groups of screaming Belgians, or a veteran's reunion.

"Do you have to anticipate a permanent excess capacity? When you add the user, the system, plus adjustment of the supply, the problem is insoluble . . .

"But there are people at the RATP who believed it was going to work. They don't know how to make the RER work, with its three missions, and they think they can make a thing like this run. 'We'll manage,' they'd say. 'We'll find a way, we'll figure it out, with advances in computer science' . . ."

"So for you, there is no mystery, because it's technologically infeasible?"

"The problem, let me tell you, is that Aramis is *a false invention, a false innovation.* The PRT has been an infeasible idea in terms of operational viability from day one." [no. 33]

---

*As every foreign tourist trying to catch a plan at Roissy or Orly must have discovered, the signs identifying the trains are mysterious: they are not names of destinations, but esoteric codes like YETI, BAYA, AZUR.

They sound awfully sure of themselves, those two. Aramis is intellectually inconceivable. It's a contradiction in terms. They're sure that it doesn't work, just by looking at operational issues, yet they put themselves in the best-case scenario. What everyone else perceives as the biggest problem of all is one that doesn't faze them at all: the 600 mobile units rolling along without a hitch.

[INTERVIEW EXCERPTS]

M. Etienne, from Matra:

"But was it perfected? Could you have gone from a few minutes' running time to the months anticipated in the contract?"

"By November 1987, we had finished integrating all the most complex functions. We needed a year, a year and a half, to make everything *reliable*—to go from minutes to months, as you say.

"Everything had to be debugged; there is no reason to think that we wouldn't have brought it off. We would have produced the contractual Aramis, and even the nominal Aramis. *No doubt about it on our end, no doubt at the RATP, no doubt in the Institute.* But it would have been especially costly for Matra, since that phase wasn't well financed; the debugging hadn't really been anticipated. No, it was feasible at the upper economic limit of what had been anticipated five years earlier."

"Is it comparable to the Concorde?"

"It exasperates me to see it compared to the Concorde. The problem is the market; if they'd built 500 Concordes instead of 20, it would have worked. Since everything is new, whether it's the TGV, the Concorde, or Aramis, it becomes economical only if there's a huge outlet for it. *It's a question of voluntarism,* the determination to continue working on Aramis did not entail a determination to continue with the line." [no. 21]

Ow! Even over the Concorde there are controversies. Now it's a question of continuity in the political will. Everything new has to be imposed by will. It's worst of all when will is subject to eclipses, *volontas interrupta,* as my mentor would surely say. They push the Concorde prototype, which is efficient and perfected, but not the 500 Concordes that would, as a bonus, make the first one profitable.

Parenthetically, M. Etienne doesn't agree with his collaborators. They say that they'd have pulled it off if they could have simplified. They don't claim they would have fulfilled the contract.

[INTERVIEW EXCERPTS]

In the Transportation Ministry office that deals with new research projects, M. Hector is speaking. He oversaw the Aramis operation even though Aramis was not viewed as a research project:

"I remember, I was there at the very beginning of Aramis. I had both of the project heads in front of me, Cohen and Frèque, with two fallback projects; one was the head of Aramis, the other head of VAL.

"Maybe the problems with VAL were overcome because the system was simpler, more realistic—enough so to convince Lille and the urban community. We never found that with Aramis, because we never had the technical proof that it would work. The increased cost has to be justified, for cities, for financial backers, *by technological performance.*"

"How do you analyze the technological feasibility of the project? Did the problem of intrinsic security that Cohen was so insistent about play a role?"

"No. I think the system simply didn't succeed in overcoming the technological difficulties well enough, fast enough, to attract financial backing easily."

"But you're aware that your interpretation is not the usual one? The consensus is that it's technologically feasible but that there's no market."

"Yes, I'm aware of that, but it must still be said that the technicians were unable to deal with the problems in time; it's absurd to say that *it's technologically perfected when the CET program wasn't carried out successfully.*

"I have to say that there's something a little phony about all of us; not you, because you're not in the field, but the rest of us. If we hear someone say, 'It's a failure,' we say, 'Oh no, it's all been very useful, there have been advances, it helped with VAL.' There have been advances, I don't doubt that. But I can't say that the system as a whole has been a technological success that advanced us toward the intended goal.

"Well, you have to know what you want. If you wanted a system, and they get a few crumbs out of some of its components, you can't call it a success, *in any case not a technological success.*

"Either they didn't put enough energy into it—but I don't think that was the problem—or, quite simply, *when the program doesn't work well technologically, it becomes harder and harder to find money.* No, no, when they tell you that 'it's technologically perfected, but there's no market,' it doesn't wash. If they'd really solved the problems, really resolved the technological difficulties, *then they would have found financing.* At least they would have found it more easily; they would have at least fulfilled the contract.

"The heart of the problem, as I see it, is that Matra was looking for improvements to VAL and never did believe in Aramis." [no. 31]

So it's technologically infeasible after all. I'm back to my original interpretation. Aramis is not viable: not only is it operationally unviable, but the mobile unit itself is impossible to build. As for the secondary benefits, they're a justification after the fact, a rationalization, for saying that the whole thing hasn't been a waste. Norbert is completely mistaken. We have to go back to doing "classic" sociology, as he calls it; we've got nothing but scandals, irrationality, coverup operations, and ex post facto rationalizations.

[INTERVIEW EXCERPTS]

M. Larédo, then with the Research Ministry, a specialist in science policy and evaluation of research programs:

"It's a problem of science policy, of project management, if you like. We tried to use the CET to do far too many things at once, things that weren't all on the same level.

"I do understand what you're asking, but as for the question of technological feasibility, *you really have to be organized to be able to answer it.* Otherwise, you don't know, and here's where the CET failed: *we can't answer that question.*

"The CET was also supposed to be the head of a line in a network, all at the same time; that was the fatal flaw.

"The Center for Technological Experimentation is a *botched compromise* between experimental development and building the head of a line. The mechanics of the transition to the CET stage *condemned the project to fail.* No, that's putting it too strongly; let's say that it ran the risk of leading the project to failure if the most optimistic possible results weren't achieved right away.

"The experimental development, in Ficheur's mind, was aimed at identifying the technological snags; everything else was secondary.

"It's easy to see how it was different from Orly. Orly was the prototype with laboratory methods; the CET is still the prototype, but with assembly-line methods. Aramis wasn't yet in production.

"The CET should have tested all the technological solutions. Half of the budget was spent on bullshit.

"The CET was conceived with a clear idea, clear to Ficheur, of *adding a stage* that was usually skipped: the stage of experimental development with industrial techniques. That stage can't be skipped without a lot of money, and not in a period of financial crisis.

"Michel Ficheur, who was with the MST (which had developed VAL), and Cohen, who was taking care of aeronautics at the Research Ministry, had clear ideas on that subject. Roughly speaking, you have five phases* that really need to be kept separate. First, there's basic research, the development concept; broadly speaking, that's the Bardet-Petit epoch. Then you have Phase 2; this is Orly, you use laboratory techniques and you verify that it's not a complete load of hogwash. But after that you have a Phase 3, and this was new, this was experimental development, not with laboratory techniques, now, but with *industrial techniques.* In the European Community they call that the pilot project. You identify the technological snags, that's all. If the thing doesn't work, you go back to Phase 2 or even to Phase 1. Then, yes, you go on to the industrial-development phase, and you get ready for homologation with a superprecise and superrigorous set of specifications. And then you go to Phase 5, the demonstration phase. In the case of VAL, this was the first segment; you get it up and running for several months and then you go on to do a series. Well, what happened with Aramis? *[He draws a sketch—see Figure 19.]*

"So according to your schema the CET, instead of covering Phase 3, covers three phases at once?

"Yes, and even four, because there are moments, in the software, in the operation, when you get the impression that some people were going back to Phase 1, to basic research, and at the same time others were settling the

*The phases and their numbers do not correspond to the RATP numbers used up to this point; the latter are assigned only in terms of budget categories corresponding to contract signatures.

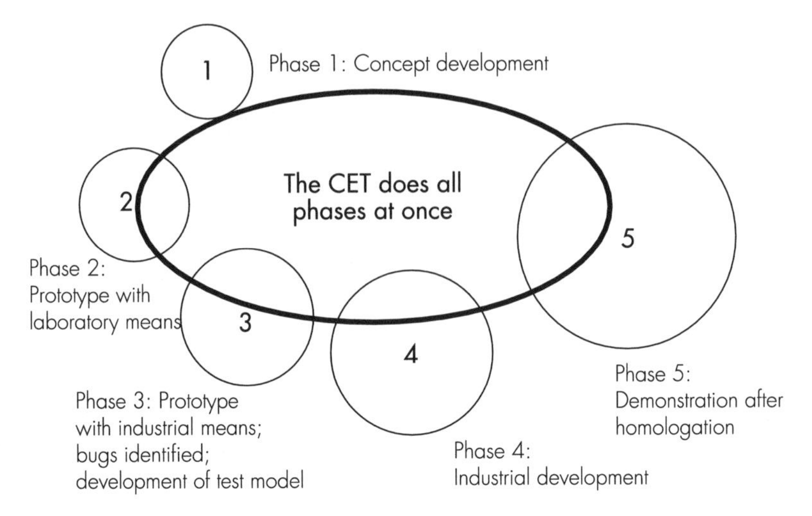

Figure 19.

question of the soap the maintenance staff was supposed to use to wash the windows. *You begin to see the tension?* You can't keep four phases going at the same time.

"And on the other hand, the RATP was dealing with the problem in Phase 4 or 5 as if it were a new metro train, while some Matra engineers were going back to first principles!

"Instead of making the technological alternatives comprehensible and comparable, *the CET locked Aramis in.*

"For Ficheur, the CET was a validation, not a homologation.

"But there was too much that wasn't clear, and anyway the RATP doesn't have a handle on that stage. The State doesn't either, by the way; it isn't able to say, 'You're going to develop and validate.' The ambiguity is there from the start. They weren't even able to identify the critical paths on which to focus their study: for example, the operating conditions were derived from those of VAL, so it was logical to put off thinking about them until later; they weren't on the critical path. They existed elsewhere and had been resolved elsewhere.

"Maintenance, on the other hand, ought to have been a problem for the CET, but it was never dealt with. It was in the original idea; it's on the critical path.

"The whole issue of stations and their design was not a problem they dealt with; they should have postponed it till later.

"Matra was always pushing to combine the phases. That's normal; it's up to the State, to public authorities, to keep them separate. For Matra, it's an incredible deal: the company gets paid for development, with subsystems that interest it; homologation serves to make the whole thing credible. So much for the Matra side.

"On the RATP side, you have the opposite problem. They ought to distinguish among phases, but they don't have the expertise. The RATP feeds on generations of metros that they improve incrementally; so for them, homologation *is very much like incremental development.* In any case, it's not in the culture of engineers.

"If people had the courage to say, 'It's risky, it's seductive, we don't know how to make it work yet, so we'd better take it easy, move ahead step by step,' that would allow projects to evolve in a very different way.

"At the RATP, the engineers don't know how to do it—it's that simple. At Matra, the engineers do know how, but commercial logic led them to suppress or repress their knowledge. So as soon as the RATP takes on the project, and it's impossible for them *not* to take it on, they apply their logic of homologation.

"They should have taken a different approach. They should have done it in the provinces. If they'd chosen Montpellier, I'm convinced that Aramis would exist today." [no. 35]

All right, fine, so now it's a problem of project management. They weren't properly organized to answer the question of technological feasibility. If the CET had been better conceived, they could have answered my questions. But I don't have 150 million francs to do it over, personally, plus three years and two hundred employees. I'm just one guy, and only an intern at that! You have to pay a staggering price to answer a question that's really pretty simple. And they keep referring me to Montpellier, which I know nothing about.

[INTERVIEW EXCERPTS]

M. Coquelet, an official with the Ile-de-France Region:

"Overall, it's a failure as a transportation system but not as research—contrary to what you're saying. Lacking more precise information, I think there were *repercussions,* for SACEM not very convincing ones, for the second-generation

VAL. You'll see, we're going to find pieces of Aramis in Matra's rockets. There were cultural *repercussions* as well: the more or less unanimous acceptance, except by the unions, of automatic piloting, and acceptance by the SNCF of the SACEM guidance system.

"No, I think you really have to *qualify the notion of total failure.*

"You should ask M. Fourcade about the reaction of the regional representatives. I myself brought the regional transportation commission to see VAL and Aramis in the winter of 1986. The visit was divided into two stages. First, we saw the vehicle and tried it out; at that point it was under manual control. *The impact was very positive.*

"Then we took them to the hangar, where there was a suspended module; we looked at the interior, the physical components. *They shrieked in horror:* 'It's a Formula One, it's monstrous, it's hyperfragile.' They began to divide up the price to see how many Rolls Royces they could buy for the money. Here, the reaction was unanimously negative." [no. 34]

Aramis, poor Aramis—now elected officials are sticking their noses in as if they really counted, poor blokes. People are telling themselves now that you didn't die in vain, because pieces of you will turn up in Matra's rockets! What's worse, now you're reduced to cultural repercussions; thanks to you, specialists in the classic metro have gotten used to the idea of automation! You were only a pretext. A red flag to distract people long enough to stick a few banderillas into the poor populace and force members of the drivers' unions to give up their trade.

[INTERVIEW EXCERPTS]

M. Pierret, elected representative from the city of Paris, in one of the elegant fake-Renaissance offices in the Hôtel de Ville:

"You have the impression that Aramis failed because in the end it didn't correspond to the city's needs?"

"No, it corresponded to a real need, a need that turns up at all levels—on the Francilienne beltway, on the peripheral boulevards. We have a radial Paris. On the Petite Ceinture, we needed to do something.

"We aren't technicians ourselves, but Aramis seemed well-adapted in terms of volume and adaptability to the site; plus it had branches to allow pinpoint service. No, we saw the project in a very favorable light."

"Later, it deteriorated?"

"Yes, in any event *we weren't kept informed;* I couldn't tell you why it didn't work out; there wasn't enough information about the difficulties.

"There was also—how can I put it?—a phenomenon of irritation after the fact.

"I'm not trying to say what really happened, just how things felt to people at the time.

"We saw the RATP build a little station, a little catwalk, a little offshoot, very well-conceived mechanisms, signal systems, lighting systems, everything very fine-tuned. It sort of gave the impression of kids having the time of their lives.

"The mayor of Paris went out there *[see Photo 15].* It seemed absurd to work on the smallest details of the platforms or the signal system when one had the impression the thing wasn't working. To my knowledge (again, I'm not trying to give you some sort of absolute truth here), the stumbling block, the real no-no, was when the RATP had to face up to the following question: 'We have to build a bridge to connect Bercy and Tolbiac.* What do we need to anticipate for this bridge? Will we have to put Aramis on it?' 'Uh, well, no, Aramis isn't going so well, don't take it into account for the bridge.'

"That's how we found out it wasn't working; aside from that, there were leaks, rumors . . ."

"So, for you, whether it's research or not doesn't bother you as long as you're told what to expect."

"No, it doesn't bother me. They haven't got it to work? That's normal, but yes, they should tell us. The probabilities need to be revised periodically, and we need time to study the alternatives.

"You understand, we're led to expect great things: it's the wonder of the century, it's France seizing the initiative once again, it's Matra, the glory of our industry, et cetera, et cetera. And then we hear no more about it, nothing more is said—suddenly, nothing. It's a mysterious thing, we're dealing with highbrows; it's government in the most disturbing sense of the word.

"You really need to do exactly the opposite. You need to *explain to people*

*Two Paris stations located on opposite sides of the Seine. The new bridge is still supposed to include a modern, lightweight, short-distance transportation system.

on an ongoing basis how things are evolving, if you want society to be more sympathetic.

"But here, with Aramis, no; it was a *panacea,* and then all at once it was a *very bad* thing. We don't know whether it's because it was a technological failure, or because the officials didn't want it, or because it was too expensive.

"I have to say that on our part—here I'm speaking on behalf of the elected officials involved—there's a certain *resignation* in the face of supreme beings; we're dealing with the *realm of the gods*. It's a subject on which people don't ask any questions, with the idea that it has cost a great deal of money and that it's too bad . . ."

"But even so, as you see it, did the officials' negative reaction play a role?"

"No, not once; as I see it, it's like the aerotrain. When it became obvious that it didn't work technologically, then the reaction of people like me, elected officials, was that the experts should have noticed much sooner that the project should have been cut . . ."

"But elected officials were often said to have been reticent about Aramis?"

"I didn't have a negative impression. Some of my colleagues had a somewhat anxious feeling, as if they'd been abandoned to the mercy of a robot. But it's like VAL—I've done a lot of skiing, the cabins didn't bother me.

"No, on the whole, we found it rather attractive. The elected officials weren't against it, and that was already a great plus; but they didn't fight for it. It's a realm where they can't fight. In any event, it was taken out of their hands. No, you can't blame Aramis on the elected officials." [no. 37]

The elected officials aren't responsible; they have no technological information except press clippings (and I know how trustworthy those are), and they weren't kept informed about anything. But the funniest part is that here is an official who is perfectly prepared to accept technological uncertainties as long as he is kept informed. This M. Pierret has a lot more good sense than many of them; he would agree with Ficheur. The point of the CET is to explore the technological snags, not to let "supreme beings" "play with their toys." M. Pierret understands the hazards of the research process better than the technicians do. Necessarily: as a politician, he's well-acquainted with long shots.

[INTERVIEW EXCERPTS]

M. Frèque at Matra:

"Still, there's one bit of fallout that I hope to hold on to: the seats designed for Aramis. They're a great success. I'm going to try to convince the RATP to keep them in the new VAL at Orly!" [no. 41]

Oh Aramis, unhappy Aramis, they're really kicking you while you're down. A seat! This is what they've made of you, what they've kept of you. Fifteen years of research for a seat! "Half a billion francs. The most expensive armchair in the history of technology." Ah, I can see the headlines of the tabloids now—if only they would pay attention to Aramis' love troubles instead of Lady Di's.

After taking this plunge into the transcripts, I wasn't much further along. With the help—precarious as it had been—of my mentor's sociology no longer available, I clung to the methods of detective novels. Like Hercule Poirot when he's stuck, I had written out a list of the most significant interpretations. They didn't converge at all. Norbert was right about that. We had found all the phases innocent, yet I was incapable of eliminating a single one of the possibilities.

1. Aramis has been perfected and will be built soon.
2. Aramis has been perfected but is too expensive for industrial construction.
3. Aramis had almost been perfected; more money and more time are all that would have been needed to complete the experimentation.
4. Aramis has been perfected and would not have been so expensive if there had been the political will to produce it on a large scale.
5. Aramis has been perfected, is very expensive, and has been abandoned politically by the ministries involved.
6. Aramis has been perfected and is very expensive, but has been abandoned politically by the local Parisian elected officials despite the support of the ministries concerned with technology.

7. Aramis has been perfected as a mobile unit, but not as an operating system.
8. Aramis has been perfected as a mobile unit, and could have been perfected as an operating system, but would have been very expensive and was abandoned politically.
9. Aramis has not been perfected as a mobile unit.
10. Aramis has not been perfected as a mobile unit because the manufacturer has abandoned it in favor of its elder brother VAL.
11. Aramis has not been perfected as a mobile unit because the operating agency set impossible requirements instead of simplifying the system.
12. Aramis could not be perfected as a mobile unit, even if the operating agency agreed to simplify, because then it would no longer be Aramis but a mini-VAL.
13. Aramis simplified, transformed, transported out of the Paris region—to Montpellier, for example—could have become technologically and politically feasible.
14. Aramis cannot be perfected as a mobile unit because it is infeasible with more than three cars.
15. Aramis cannot be perfected, but pieces of Aramis have been perfected and have repercussions for other activities.
16. No piece of Aramis has been perfected; everything would have to be started from scratch if the project were taken up again. But there are cultural repercussions: Paris has grown accustomed to automation.
17. No piece of Aramis has been perfected. There are no repercussions; it is a false innovation.
18. All the questions about repercussions and technological feasibility and profitability could have been answered if the CET had been well conceived.
19. It is impossible to judge. The question of the technological possibility of Aramis is a black box.
20. The question of the technological possibility of Aramis must not be raised, so that the Budget Office will not keep coming around harassing guided-transportation systems.
21. The question of technological feasibility will not be raised.

My list went from the clearest points to the most opaque. From lightest to darkest. From most open to most secret. And not a single element was stable. Either Aramis really existed and it had been killed (the elected officials, the Budget Office, the politicians had killed it; there really had been murder, blindness, obscurantism), or else, at the other extreme, Aramis had never existed: it had remained inconceivable since 1981, and a different crime had been committed by a different sort of blindness, another obscurantism; for years on end they'd been drawing funds for nothing—a pure loss. In the first case, I was explaining Aramis' unjust death in 1987; in the second, I was explaining Aramis' unjustified reprise in 1984. But there's the rub: I couldn't explain them both at once, and I couldn't choose one over the other. Norbert was right: I was like Buridan's donkey; I was going around in circles, indignant at having to do my professor's work, but furious at being unable—me, an engineer!—to pull it off better than he could.

While I was rereading, for the tenth time, the report on the end of the CET from October 1987, I finally found the hidden staircase. "Good Lord, but of course! That's it!" I exclaimed, just like Hercule Poirot.

[DOCUMENT: DETAILED DESCRIPTION OF THE AUTOMATED COMPONENTS OF THE ARAMIS SYSTEM; MATRA, OCTOBER 15, 1987; ORIGINAL EMPHASIS]

2.1. Basic Principles of the Aramis System

The principle of trains of variable length makes it possible:

—to adapt the length of the train easily to the demand for transportation. It is possible to retain a high quality of service during slack periods by means of short trains used at intervals that remain brief;

—to exploit a network with multiple origins and destinations without requiring passengers to transfer, as the trains demerge and merge on both sides of the switching points; this type of operation makes it possible, in particular, to retain brief intervals on the various branches of a network, to the extent that these intervals can be equal to the interval on the common trunk line;

—to offer, in the most sophisticated version of the Aramis system, direct or semidirect services, by using off-line stations; certain pairs of cars from a train may in effect short-circuit the station by taking the main track and avoid intermediate stops.

In addition to the specific features described above, the chief characteristics of the system are the following:

—*the small size* and thus the ease of insertion in an urban site, the minimum curve radius being 10 meters without passengers at the terminus, 25 meters with passengers on the lines;

—*the very brief service interval.*

The report presented the 1987 Aramis, word for word, as identical to Petit and Bardet's 1970 Aramis. I myself had found twenty-one interpretations, but the technological documents remained mute about this dispersion. Aramis had not incorporated any of the transformations of its environment. It had remained purely an object, a pure object. Remote from the social arena, remote from history; intact. This was surely it, the hidden staircase Norbert was looking for. Its soul and its body, as he would say, never merged.

I saw my professor again only on the eve of the debriefing session that he called "restitution." While I was stamping my feet in excitement, he seemed to be at a low point.

"Oh, I'm all washed up," he told me. "I'm going to change careers. Technology isn't for me. Even in dreams I haven't been able to embrace Frankenstein's creature. I've pulled away in horror. I've been a coward. I haven't reached any credible conclusion. I'm going to go back to classical culture. Do theology again. Reread Tacitus, as my Polytechnique colleague Finkielkraut advises. Habermas is right, after all; there's no love, no culture, in technologies. The farther behind you leave them, the better you think. How about you, what have you found? We do have to turn in a report. And you're the engineer, after all."

"I was! I could have been! I was going to be! And you're the one who dragged me into this 'mission impossible.' And you're asking me to pass judgment in your place? It's your job to present the conclusions."

I showed him the 1987 document.

"But we've read that a hundred times," he said dejectedly. "They've been saying the same things for fifteen years, ever since Petit and Bardet. Nonmaterial couplings, small size, adaptability to the sites, no transfers, comfort for seated passengers, a network with multiple origins and destinations. It's all there. Where does it get us?"

"But that's just the point—it doesn't get us anywhere. Look at the date: October 1987, one month before Aramis' death. Aramis has been exactly the same for seventeen years. The basic concept hasn't undergone any transformation, any negotiation, except for the pair of cars and the ten seats. It's held up against all comers. Yet you interviewed quite a few skeptics! Things have happened in the last fifteen years! And now, look at my list: the interpretations are all over the map."

His eyes lit up when he had run down the list.

"You see," I continued triumphantly. "Nothing changed it. It didn't incorporate any skepticism, any random event. It reaches the moment of death absolutely intact, fresh as the day it was born. Without aging, without being 'degraded,' without being 'adulterated,' as they all say. On the one hand, Aramis; on the other, my little chart."

"But wait, I don't understand," Norbert interrupted, more and more intrigued. "There's absolutely no relationship between these two documents."

"Exactly. Does Aramis absorb the 600 pairs of cars?"

"No, operational viability comes later, at the end of the road."

"Do they transport it to Montpellier?"

"Ah! No, everybody talks about it, but it remains a pure possibility."

"Do they increase its height so as to be able to include standing passengers and restore elasticity to the traffic flow?"

"Oh, no, that's at the very end, and it isn't Aramis any more if people aren't seated."

"Are they interested in elected officials, in what they think?"

"No, you can't expect too much from them. They're dragged into it only during the final months, and they aren't happy."

"Do they take into account the skeptics at Matra or at the RATP?"

"No, not at all."

"Well then, there you have it, Norbert! They don't discuss it. They don't know what research is. They think it amounts to throwing money out the window! While everything is shifting around inside the Aramis mobile unit, outside everything is carved in stone. They don't renegotiate. The only one who explicitly accepts research as such is an elected official who has nothing to say!"

"I'm beginning to understand. Yes, yes, there really is love in technologies. Poor Habermas! For a minute there I almost agreed with him. That's where the formal defect was, the sin, the crime—right in this chart. Your chart. They didn't make Aramis a research project. They didn't love it. You've saved me."

We hugged each other.

"But then we have to understand," he continued, as excited now as I, "whether our interlocutors actually do hate research. We have to go back to the interview transcripts."

[INTERVIEW EXCERPTS]

M. Piébeau, technical advisor at the Transportation Ministry:

"Were there warning signals during the course of the CET? Did you have the impression that it wasn't working?"

"Zero. Not a word, not a single red light. Well, all right, the time frames were slipping; but after all, that's already pretty normal in industry, so what can you expect in *state-of-the-art research*? Everybody thought things were going quite well.

"Except of course from the operational standpoint. When the report on operations came out, then people started asking questions. They noticed that managing 600 pairs of cars was not so simple; just getting them cleaned was a real problem.

"You talk about 'state-of-the-art research,' but, excuse me, the research part

was supposed to be finished in 1982. They were heading toward homologation, production."

"Yes, but it's the repercussions that count, after all."

"Forgive me for insisting on this, but people thought about repercussions after *the project was halted. There isn't any document that talks about research or repercussions before the end of 1987."*

"Yes, of course, but there they thought they were building a line, I mean before. You do have to talk about lines; otherwise, in transportation, *there's no money for research.* People *don't like research*—especially the Finance Ministry, obviously."

"Again, I apologize, but this is really the crucial point. What about you, did you think it was research or production?"

"I'd be tempted to say that, before, I really believed in the transportation system, even if I had some doubts about the technology; but *afterward, yes, it was closer to research,* and besides, technologically, in the end, it worked, *even though it hadn't been perfected.* But the repercussions are important; it's a good thing, too, because there are already enough skeletons like that in the closets." [no. 13]

M. Maire, RATP's technical director responsible for innovations:

"It hasn't been proved in any case *that it was impossible . . ."*

"You've been criticized, as you know, for refusing to do a mini-VAL, for refusing to simplify?"

"Those people, the ones who say that, underestimate us. We chose Aramis because it was more complex, more high-tech, even though our relations with Matra were pretty complicated.

"Aramis is a system *that is meaningless unless it remains pure.* Now all the PRT systems have gotten weighed down and have lost their point-to-point capacity. If it's *watered down,* it isn't interesting.

"So, for example, with nonmaterial coupling, we thought that would be a useful side effect, but *it wasn't our goal.*

"The setup was flawed industrially, defective from the start. Matra wanted useful fallout right away, while we knew that it wouldn't happen right away—it never does in *advanced research."*

"So then it is research?"

"No *[impatiently],* we were aiming at the main goal, not for the *fallout.* The outcome went well; at the administrative council of the RATP everybody reacted well: 'It's not a technological failure; there will be useful fallout.' They emphasized the repercussions in order to avoid sounding negative."

"But then why not say it was research?"

[More and more annoyed] "No, really, *it isn't research,* having a CET—it's more instructive. We could have lined up research projects without doing a thing. Orly was much less convincing than the full-scale trial. With research, *you're throwing money* out the window, you're *going every whichway.*

"No, no, I'm a partisan of the full trial. You have to *finalize;* that forces people to get specific about their ideas. It has to work on the site; I really believe in the value of the concrete realization . . ."

"Excuse me for dwelling on this, but this is the essential point. Finally, in 1988, the overall project is justified by the repercussions, though people don't know exactly what these are, which is normal if you're doing basic research; but when I tell you it's research you say no, it's finalized, it's full scale?"

"But how do you expect to finance a research budget like that? It's impossible; it's not aeronautics, or nuclear power. It really does have to be finalized, reusable; there has to be a line."

"But in the end, there isn't any line, or any research either."

"Of course there is! There's the fallout: it's irrigating the entire world of transportation, and it's been highly profitable to Matra—to us, too, in the end." [no. 22, pp. 15–20]

M. Gontran, technical advisor in Fiterman's cabinet:

"But you knew it was state-of-the-art, hyperrefined technologically, that it looked more like research and that consequently, if you wanted useful fallout, it was crucial not to finalize, not to shut down too quickly?"

"There's the whole problem in the relationship between technicians and politicians.

"Politicians skim over this sort of detail as soon as the DTT or the RATP says the thing can work. If the press has got hold of it as well, there's a real stampede. It's the logic of the media.

"Okay, it's true, I didn't oppose it. I made notes. I said, 'This part and that one haven't been tested,' but that sort of doubt is perceived, by politicians, as typical of researchers having a good time, nitpicking; it's perceived as perfectionism. If it's a question of safety, then yes, that interests them, but the rest, no."

"A research project can't be sold as a research project? Why does it always have to be packaged?"

"Because it's politically unmanageable. Because of the announcement effect, you can't sell a project and ask for five years. The central administration *can't understand the research process.* You can't want to go back over things that have already been achieved. *Research requires too much time.*

"It's an atmosphere of *generalized positivism*. Science has an answer for everything, and a quick one. There are no *direct connections between researchers and politicians*. This is a big problem. The DTT played the role of a *screen*. Since it's aware that it can't manage the project technologically, it passes it over to the RATP." [no. 42]

M. Rescher, director of ground transportation in the ministry at the time of the interview:

"A research-and-development project on this scale, on a very specific object, and with uncertain prospects for commercialization—there's no way it can get financing. So the idea was to spread out the burden of financing between the Region and the manufacturers, and for that they had to say they intended to build a line. It's clear that for the Region, research isn't its mission; as for the manufacturers, well, it's normal, they expected useful fallout in terms of production."

"But why didn't the uncertainty about the existence of the line have a negative impact on the CET? On its missions?"

"You have a point, but we were always told that the line and the three billion francs to finance it weren't guaranteed at all.

"The doubts about the technological aspects never got to me; we were told that the motor 'works.'

"You're right *about the ambiguity between research and development;* there you're touching on a real problem. But you can't put 150 million francs into it if you don't say that it can be integrated with the beginning of a line, with a real public-transportation system.

"At the same time, it's true that it's not very logical, since in the phasing they shouldn't have started with the boulevard Victor."

"But then, why not spend much less on the CET, and really test Aramis?"

"Well, because you need to promise a line. Again, without that you don't have the Region behind you; you probably don't have Matra."

"But since, in the end, you don't have a line at all . . .?"

"Yes, I see the problem; but if it's too far ahead of its time, does that motivate people? Certainly not the Region.

"I think there's another problem: it's that we at the DTT weren't informed about the technological concerns. Government oversight of businesses and operating agencies is not cut-and-dried. The technicians push the project ahead and they take a certain pride in not letting the problems reach us, but *we can't put a cop behind every researcher.*

"As far as we're concerned, *all the doubts arose* after the codicil, here, when we talked about it. The elected officials didn't want it, the economic studies were bad, operation was becoming impossible, technologically it was difficult—but all that was *after* the codicil. From 1984 to 1987, as far as I could tell, there were no obvious problems except some delays, but that's normal." [no. 38]

M. Antoine, at the RATP, on the executive floor:

"If I understand correctly, you supported the CET but not Aramis?"

"Yes, you could put it that way. For me, as an engineer, getting the rotary motor to exceed 30 kilowatts, that's terrific. If it worked, there'd be no more reduction gears, no more elasticity—it'd be great. For the time being, we don't know how to do it, but maybe someday we will.

"And electronic coupling is the same story. If we'd done the development, in the CET, it would be colossal, on the European scale, and we'd have said, 'Here's the development, now we have to keep going' . . .

"I always told Matra: 'There's just one thing that interests me, and that's electronic coupling, if it's cheaper than the mechanical version.' Aramis would have made its contribution to the *research* on lowering materials costs; that's what I was hoping for from the CET.

"If it had been up to me, I would have preferred to spend 100 million francs with all the builders working together. I would have said: 'We've made VAL, the first generation, now we have to shift over to the second-generation VAL, which should have the same capabilities but cost 30 percent less.' That's what I would have done in 1984, but they didn't ask my advice.

"There was some useful fallout, though—otherwise it wouldn't bear thinking about.

"Automated mechanisms ought to evolve according to *modular* concepts, like Lego, in the early stages, and then in *standardized* black boxes. I was expecting they'd be able to pull a speed-control module out of Aramis, for example. For the time being, they're bringing a specific solution to every specific problem."

"But why not turn the CET into an R&D project, then? The modular approach is entirely different from trying to build a complete system."

"Because it's impossible. The industrial milieu is a battleground. They pick away at each other, they're like roosters on a pile of manure pulling out each other's feathers, and pretty soon they won't have any feathers left! . . ."

"I still don't understand: since it is research, why not say so?"

"Because the CET was tied up with the hope of constructing a line and with a prudent approach on the part of the RATP, which was both enthusiastic and cautious.

"They couldn't sell the project except by saying it was feasible; this was *packaging*, to come up with adequate funding."

"But by doing modular research, they would have spent less, they would have interested more people, and they wouldn't have left the impression that they'd failed, they could really say there was useful fallout. Whereas now, the fallout argument gives the impression that it's, how can I put it, a rationalization."

[Long silence.] "I do see the stumbling block. Yes, it does call into question the decisionmaking process."

"There wasn't a lot of discussion."

"It was a very deliberate action on the part of the general director and the president. I have to say that *there was no discussion* among the people who had things to contribute.

"I can't be much clearer than that.

"Skeptics were considered retrograde. *It's hard to argue in a situation like that."* [no. 36]

"Well, now we have the key to the puzzle," I exclaimed enthusiastically; "the fallout from advanced basic research always comes after the failure of the project! The whole thing should have been a research project. They abandoned technology while thinking that it was going to be finalized all by itself, that it was autonomous, that they'd see how things worked out afterward, that it had to be protected from its environment."

"Yes. They really succeeded in separating technology from the social arena! They really believe in the total difference between the two. To cap it off, they themselves, the engineers and the technologists, believe what philosophers of technology say about technology! And in addition, research for them is impossible, unthinkable; its very movement of negotiation, of uncertainty, scandalizes them. They throw money out the window, but they think research means throwing money out the window."

"Not enough negotiation!"

“No, no, not enough love! Love and research—it’s the same movement. They abandoned Aramis so as not to compromise it; they committed the only sin that counts—the sin of disincarnation. They’re hardened positivists; they believe the soul and the body are distinct,” murmured my mentor, completely enraptured.

EPILOGUE: ARAMIS UNLOVED

Quai des Grands-Augustins: all the major players in the Aramis affair are seated around a large oval table. The project heads from Matra, the RATP, the Region, the research institutes, and the government ministries have all been convoked by the clients of the study. Only the elected officials are missing.

"In detective stories there is always a moment when all the suspects and their buddies gather in a big circle, quaking, to hear Inspector Columbo or Hercule Poirot name the perpetrator," Norbert began.

"The guilty party is often the one who is the most ill at ease, who says foolish things and gives himself away. Yet even though our situation this evening bears some likeness to that one, even though you've been summoned here, we're not in a detective story, for two reasons.

"First, because today, around this table, the person doing the most quaking, the one who's going to say the most foolish things, who's going to give himself away, is me. Don't go looking for any other perpetrator.

"No, I'm not Hercule Poirot—I'm not going to reveal the truth, unveil the guilty party, or unmask anyone. We get the truth only in novels, and this isn't a novel. In real life, reality sets anyone who looks for it to quaking all over.

"It isn't easy to restitute what killed Aramis in this company, under your watchful eyes, knowing that you have often given me information about each other in strict confidence. And I must betray no secrets, yet you are expecting me to uncover the secret.

"Don't worry, I can no longer escape.

"But sociologists are often noted for being good at getting themselves off the hook. They practice a scorched-earth policy and then disappear as soon as the interviews are over; they go off to talk to their peers, other sociologists; they don't give a damn about what their 'informants,' as they put it, will say; and anyway the informants have to be forgiven—they don't know what they're doing.

"But today, it is to you that I am making restitution, and what I am restituting is what you have told me. What I have to be forgiven for is everything I know, everything you've said to me. You are the ones who trained me; you are also my judges. You are the ones before whom I have to make myself understood, because I have tried to understand you. You are the ones I do not want to betray, yet you are expecting me to pass judgment. You are quaking—less than I am, of course, but still, we are even, because you are the judges of my judgment as I am of yours.

"The second reason that what we are doing here cannot be equated with the final chapter of a detective story is that Aramis was not murdered.

"Aramis is dead, but there was no murder. There is no perpetrator, no guilty party. There is no particular scandal in the Aramis affair. The funds expended and the time frame are normal. There has been no scandal. The Personal Rapid Transit concept is somewhat out of fashion, but failing to follow fashion is not a crime, and tomorrow people will no doubt be talking about PRTs again. The Petite Ceinture line needed to be equipped; it still needs something on its tracks; tomorrow it will probably be outfitted with something that will look very much like Aramis, at least in its dimensions. You have all told me the same thing, one way or another: every part of Aramis was necessary; each one still corresponds to a need and will most likely resurface later. No, there's nothing scandalous here. Everybody, every one of you, believed you were doing the right thing. There wasn't a shred of wickedness in this collective drift of good intentions. Even the Machiavellianism—and there was a bit of that—was not pursued vigorously enough to let us designate a mastermind or a bad guy. Lined up together, all the accusations—and there were many, some of them pretty potent—cancel each other out. From the standpoint of justice—even from the standpoint of the immanent justice that has you all sweating a little, that of the Budget Office—there would be a finding of no cause.

"Yet the good intentions drifted. Aramis died in 1987, Aramis came back to life in 1984, it didn't take place, it did take place.

"We are gathered this evening around Aramis, around an object that did not take place but that was not without object. The proof is that we all loved it. I myself am an outsider in the world of guided transportation, and I too have been susceptible to the contagion. I don't mind telling you all that I really loved it. I shed real tears as I followed the ups and downs of this being that asked to exist, to whom you offered existence.

"If Aramis didn't exist, it would have to be invented. If Aramis had not tried to exist, it would have had to be invented. If you had not tried to invent it, you would have failed in your mission. Anyone who does not feel this every single day, taking the metro or getting into an automobile or swearing in a traffic jam, is not part of our circle, has no claim on our attention.

"Yet we were wrong, we made mistakes, we misled one another. Where is the error? Where is the crime? Maybe the question needs to be put differently: Where is the sin?

"Aramis has been fragile from the outset—we all know that; not fragile in just one respect, in one weak link, as with other innovations, but fragile on all points. It is limited—"hyperrefined," as you put it. The demand for it is undefined, the feasibility of the vehicle is uncertain, its costs are variable, its operating conditions are chancy, its political support—like all political support—is inconsistent. It innovates in all respects at once—motor, casing, tracks, chips, site, hyperfrequencies, doors, signal systems, passenger behavior. And beyond all this, it is hypersensitive to variations in its environment. A case of shilly-shallying: the history of Aramis is proof. Not one ministerial portfolio has changed hands without Aramis' coughing and catching cold.

"Yet in spite of its fragility, its sensitivity, how have we treated it?

"Like an uncomplicated development project that could unfold in successive phases from the drawing boards to a metro system that would run with 14,000 passengers an hour in the south Paris region every day, twenty-four hours a day.

"Here is our mistake, one we all made, the only one we made. You had a hypersensitive project, and you treated it as if you could get it through under its own steam. But you weren't nuclear power, you weren't the army; you weren't able to make the ministries, the Budget Office, or the passengers behave in such a way as to adapt themselves to Aramis' subtle variations, to its hesitations and its moods. And you left Aramis to cope under its own steam when it was actually weak and fragile. You believed in the autonomy of technology.

"If the Budget Office can kill Aramis, what should you do, if you really care about it? Impose yourselves on the Budget Office, force it to accept Aramis. You can't do that? Then don't ask Aramis to be capable of doing it on its own. If elected officials from the south Paris region can kill Aramis, what should you do? Make them change their minds, or get other ones elected. You don't think you have the power? Then don't expect that Aramis will. The laws of physics and the three-body problem make it impossible to calculate the displacements of more than three pairs of cars? Can you change physics, redo Poincaré's calculations? No? Then don't charge Aramis' pairs of cars with the overwhelming task of handling, all by themselves, knowledge that even God Almighty may not have. It's only in horror stories and epistemological treatises that omniscient humanoids invade the world on their own and reshape it to their own needs. In worlds like theirs, don't ask the impossible. You want Aramis to be automated, irreversible, real? It will be, it could be, it could have been. But at the beginning it is still unreal, reversible, manual, terribly manual. Don't ask Aramis, don't ask a project, to do something you, as individuals and corporate bodies, find yourselves incapable of accomplishing. Either you change the world to adapt it to the nominal Aramis, or else, yes, you need—you needed—to change Aramis.

"But then you would have needed to acknowledge that this was a research project.

"Oh, you do love science! You were formed by it in your graduate schools. As for technologies, you drank in its certainties with your mother's milk. But you still don't love research. Its uncertainties, its whirlwinds, its mixed character, its setbacks, its negotiations, its compromises—you turn all that over to politicians, journalists, union leaders, sociologists, writers, and literary critics: to me and people like me. Research, for you, is the tub of the Danaïdes: it's discussion leading nowhere, it's a dancer in a tutu, it's democracy. But technological research is the exact opposite of science, the exact opposite of technology.

"Ah! If only Aramis could speak! If it were Aramis speaking to you here instead of me, if it were he who had called you together, gathered you around himself, he would have plenty to say about those changes you didn't ask him for in order, so you said, to respect him, to keep him pure."

Yes, he'd say to M. Etienne, who is here today, *you modified me from top to bottom, but to turn me into what? A mini-VAL, a more compact version of my elder brother VAL. I was willing, I would have been quite prepared to come into existence in that form; but then you, M. Maire, who are here today, you hated me, you accused M. Etienne of degrading me, of adulterating me. I interested you only if I remained complex, I had to have all my electronic assets, I had to be able to couple and uncouple electronically. I would have been happy to do that; perhaps I could have. But you loved me then in separate pieces, pieces that could be used again elsewhere, to help out your business, I don't know where, in the RER, in Lyon, at Orly. You loved me provided that I did not exist as a whole. And then you, M. Coquelet, you who represent the Region: if I did not exist as a whole, you wanted to hear no more about me, you threatened to have nothing more to do with me—you and your billions of francs and your millions of passengers. Then people grew frantic on my account. They had meetings about me again. I had to exist as a line so everybody would still love me, so the Budget Office would support me. I would have been happy to be a line. What more could I ask of the god of guided transportation? But then what a fuss there was, in your own services! The engineers threw up their hands. M. Frèque, who is here today, said I was impossible, and you, M. Grinevald, you called me infeasible; you even accused me, I remember, of being a subsonic Concorde, inoperable, unprofitable, and passengerless, or, I'm not sure, maybe an attraction for an amusement park. What? I could have existed, for one of you, but on the condition that I should exist as a prototype and not transport anyone! I would have been happy to be something, in the end, anything at all—but first you have to agree among yourselves. I can't be everything to everybody. The finest project in the world can't give more than it has, and what it has is what you give it.*

"That is what Aramis would say, and perhaps he'd be even less tactful than I. Perhaps he'd forget himself and accuse me in turn of faintheartedness."

You're really too gutless with them, he would cry out suddenly like a condemned man who challenges his lawyer and addresses himself indignantly to the court. You're in cahoots with them. I have no use for your sweet-talking ecumenicism. Not to blame anyone—that's too easy. No, no, he would perhaps say, a terrible doubt has struck fear into my heart time after time. If they cannot reach agreement on my behalf, if they refuse to negotiate with one another over what I am supposed to be, it's because they want me to stay in limbo forever. For them, I'm just something to talk about. A pretext-object. One of those plans that gets passed around for years so long as they don't really exist. No, no, you didn't love me. You loved me as an idea. You loved me as long as I was vague. The proof is that you didn't even agree as to whether I am possible in principle, whether my essence does or does not imply my existence. Even that would be enough for me. Oh, how happy I would be to return to limbo if I knew that I was at least conceivable. I won't be granted even that much.

You built the CET by mistake, to salve your consciences, to assuage your guilt; there had been so much talk about me for so long, I really needed to exist, to move into action. But in reality no one, during those years, could hold on to any trace of the reasons for producing me. You got yourselves all mixed up in your goals and strategies. Of what ends am I the means? Tell me! You hid from one another in order not to admit that you didn't want me. You built the CET the way human couples produce one child after another when they're about to divorce, trying to patch things up. What horrible hypocrisy, entrusting to the whimperings of the most fragile of beings the responsibility for keeping together creatures that are much stronger than itself.

Whatever you do (you say) don't argue! Don't doubt! Don't negotiate! Don't fight! But that's not how one loves brothers of my sort. Silence is for me to bear, not you. You humans need to talk, argue, get mad, that's your role in this imperfect world. A frightening conspiracy of silence is what imposed silence on me. I would exist, on the contrary, if you had spoken, you silent ones. And the funniest thing of all is that you really thought you'd said enough about me. You really had the impression that twenty years of discussion, of plans and counterplans, were enough, that it was time to quit, time to move on at last to the serious things; you had the impression that I had to be finished. But that is precisely what finished me off. No, no, you didn't argue for seventeen years, since you didn't redo me, didn't redesign me from head to toe. You skirted the issue, you concerned yourselves only with two mobile units, not with all of me. Does God abandon his creatures when they are still of unbaked clay? And even if you don't believe in God, does nature abandon its lineages in the sketchy state in which fossils are found? Isn't Darwin right? Isn't creation continuous?

It was so I wouldn't be degraded! he will perhaps shout out with a sardonic laugh.

They wanted to keep me pure of all compromise! "Be suspicious of purity, it's the vitriol of the soul." They wanted to keep me nominal, as they put it. Noumenal, rather. Well, too bad for them, since because of that insistence on purity, what am I? Nothing but a name! And what a name, by the way! How could they stick me with the name of that mustachioed swashbuckler?

But no, they didn't want to complicate their lives. Everything in its time. Later for the crowd problems; later for the social problems; later for the operating problems. There'll be plenty of time to find a place to use this stuff. As if I were rootless! As if I were a thing! As if things were things! Let's not lose time on complicated problems, you said! But didn't you really lose time, in the long run? They wanted to concentrate on the components, the motor, the casing, one mobile unit, then two, then three. But that's not how we exist, we beings made of things. That's not what brought my elder brother VAL into existence. "Technologically perfected," you say. But how do you dare treat me that way, when I don't exist? How can you say of something that it is perfected, achieved, finished, technologically impeccable, when it does not have being! As if existence were additional, as if it were supplementary, accidental, added from without to beings of reason! As if breath fell miraculously onto our bodies of clay! Accursed heretics who thus curse your own bodies full of souls, your own incarnate God, and your own Darwinian nature in perpetual agitation. But you yourselves would not exist, at that rate!

We are not little bits added one to another while waiting for a totality to come from elsewhere. We are not without humanity. We are not. We are—ah! what are we? Whirlwinds, great loops of retroactions, troubled crowds, searching, restless, critical, unstable, complex, yes, vast collectives. They wanted simple, clear, technological solutions. But we technological objects have nothing technological about us. Are all you engineers ready to hear me sputter with rage one last time, before I disappear forever into the void from which I could have been saved? You hate us; you hate technologies . . . A local elected official knows more about research, about uncertainty, about negotiation, than all you so-called technicians do. This message is veiled from the scientists and the literati, and revealed to the meek and the poor! They say they love me and don't want to search for me! They say they love technology and they don't want to be researchers! They say they love nonhumans and they don't love humans! And then there are the others, who say they love humans and who don't love us, us machines! Oh, you really do live in Erewhon, you live among things and you think you remain among yourselves. Well then, may you perish like the residents of Erewhon, along with the object of your hatred!

"But no, no, if Aramis had been able to summon you to his side," Norbert went on, enraptured, "if Aramis formed the center of a great unanimous circle, he would not speak in order to point out where you went astray. He would not speak at all. Because then he would exist! As Samuel Butler says in the *Book of Machines,* 'Won't it be the glory of machines that they can do without the great gift of speech? Someone has said that silence is a virtue that makes us agreeable to our fellows.' Ah, Aramis, you would finally enjoy that silence. Why would you waste your time speaking? What you aspire to is not bearing the 'I.' On the contrary, your dignity, your virtue, your glory, lie in being a 'one.' And it is this silence, this happy anonymity, this depth, this heaviness, this humanity, that we have denied you. I am speaking in your place, I am offering you the awkward detour of a prosopopoeia, but it is precisely because you are dead forever. 'It' wanted to become not the subject of our discourse, but the object, the tender anonymous object by means of which we would travel in Paris. Is that so hard to understand? It wanted the happy fate of VAL, its 'elder brother,' as it naïvely put it. It wanted to be silence, and thing, and object, and to spread throughout its great, finally mute body the flow of our displacements. And you, by not speaking more among yourselves, and I, by speaking so much of it, have turned it forever into a being of reason, the pitiful hero of an experimental novel. Neither autonomy nor independence. Thus, it does not exist, since it is speaking here, since it can speak through my mouth, instead of being over there on the boulevard Victor, a happy thing."

And Norbert sat back down to stunned silence.

"Ahem, ahem, thank you very much, Professor H.," coughed the presiding official who had ordered the study. "I must say, it's a real novel you've done for us. I suppose there are reactions, questions. . . Yes, M. Etienne?"

The discussion lasted until I was asked to present, more prosaically, the practical solutions that we had agreed to recommend to the RATP.

At the end of my internship, in June 1988, I met Norbert for the last time.

"What are you going to do now?" I asked him.

"I'd really like to publish that story, since everybody tells me it's a novel."

"But it's unpublishable! What about confidentiality? And besides, you didn't find the solution; you weren't able to prove that if you'd done the study five years earlier you would have seen the flaw and saved millions."

"*Obligation to use all appropriate means, not to produce results*"—that's what the fine print says in all our standard contracts. And we certainly didn't skimp on the means, I don't think. Confidentiality isn't a problem; I'm not denouncing anyone, I'm not laying any blame, there's no scandal, no wrongdoers. It's a collective drift, there were only good intentions. And I'd actually like to do a book in which there's no metalanguage, no master discourse, where you wouldn't know which is strongest, the sociological theory or the documents or the interviews or the literature or the fiction, where all these genres or regimes would be at the same level, each one interpreting the others without anybody being able to say which is judging what."

"But that's impossible; and besides, it would be incredibly boring. And what good would it do?"

"Well, it would be good for training people like you. And it would be good for educating the public, for getting people to understand, getting them to love technologies. I'd like to turn the failure of Aramis into a success, so it won't have died in vain, so . . ."

"You're funny, Norbert. You want to reeducate the whole world and you want to produce a discourse that doesn't control anyone! Readers want a line, they want mass transit, not point-to-point, not personalized cabins. You want to know what I think? You're about to embark on another Aramis project, another wild-goose chase. As infeasible as the first one. Remember the lesson of Aramis: 'Don't innovate in every respect at once.' Your book is just one more rickety endeavor, ill-conceived from birth, a white elephant."

"But if it were viable, at least it would be useful to others, to future engineers like yourself. To help them understand research."

If I had indeed discovered the importance of research. . . I was hardly convinced by Norbert's science, but since it was he, after all, who was to give me a grade on my internship, I grunted and fell silent.

He continued.

"Never mind, I'll write one more report, another colorless text, an expert's audit; I'll follow the advice of the *Times:* 'Make it boring.'"

After a few moments, he went on in a tone of feigned indifference (for I knew he wanted to keep me in his lab):

"And you, what are you going to do now?"

"I hardly know how to tell you . . . Sociology is fascinating, but I think . . ." Then I took the plunge: "I'm going to be an engineer again, a real one; I'm going to work for a big software company."

"Too bad for me, but I suppose it's good for you, it surely pays better," Norbert said in a tone at once bitter and paternal. "At least you've learned some things you can use, haven't you?"

"Uh, yes, the glorious uncertainty of technological research; but to tell the truth, in fact, I think I'm going to work now more on real technological projects, trying, really, I mean, to forget, well, not to forget, but to set aside . . .; this was after all, don't take this the wrong way, a parenthesis."

"Because Aramis . . .?"

"Yes, no, of course, but anyway, I'd like, I think, I hope to come across a technological project, purely technological, I'm not sure how to put this, but they've got a really well-conceived project, really doable."

"Ah! ah!" Norbert interjected sarcastically. "So you haven't been immunized? You think Aramis is a special case? That they could have done better? That it's pathological? I've never seen such a stubborn engineer. Aramis will have died in vain if you think it was a monster. Aramis gave you the best . . ."

"Stop, stop, no, I'm not abandoning Aramis, and you know why? On the contrary, I'm continuing it. The place where I'm going to work—you'll never guess what they're working on. A huge project to develop an intelligent car. And you know what they talk about in their documentation?" I went on proudly. "Adjustable mobile sectors, nonmaterial couplings, reconfigurable trains, ultrasound devices, UGTs! Yes, it's true—it's Aramis backward. Instead of starting with public transportation to end up making a car, they're starting with private cars and turning them into public transportation, into trains. You see, now, I'm not ungrateful! But there, at least, it's technologically perfected; they're spending billions on it."

"More than on sociology, I understand. And what's your project called?"

"Prometheus."

"Prometheus! The 'smart car'?" And my former mentor burst out laughing vengefully. "And it's perfected? And it's technologically feasible? But the stealer of fire is Aramis to the tenth power, my poor fellow. Frankenstein's monster looks like the Belvedere Apollo compared to this project."

"Not at all," I replied, piqued. "It's technologically state-of-the-art, but feasible."

"After one year! Listen to him, look what he's saying! He's forgotten nothing and learned nothing. But in five years I'll come along and study it for you, your Prometheus, my poor little engineer, they'll be asking me for another postmortem study . . ."

"In any case you won't have me as your assistant," I said stiffly. I closed the door of the office and left Norbert H. to his "refined" sociology. As for me, from now on I'd be devoting myself to hard technology.

Ah, my internship grade? A well-deserved A+.

Two years later, on the plane coming back from a colloquium on "smart cars," I was stunned to read the following article in the April 28, 1990, *San Diego Union:*

FAMILY-SIZED MASS TRANSIT CARS TO BE STUDIED AS ANSWER TO CONGESTION

"Called a 'personal rapid transit' system, the idea is to construct a network of lightweight, automated rail lines that make it possible for commuters to direct individual rail cars to a specific destination, without making intermediary stops," Franzen told reporters here.

"You would walk into a station and buy a ticket," he said. "The vehicle will read that ticket and take you exactly where you want to go.

"The technology exists all over the world," Franzen said. "It has not been put together in this form anywhere in the world . . ."

"Damn!" I said to myself. "If they'd just waited a couple more years, Aramis would have been on the right path, technologically! 'This revolutionary transportation system is soon going to transform the city of Chicago . . . Thereby solving the problems of congestion, . . . pollution.' But it's Bardet, it's Petit all over again! . . . 'A billion dollars.' They should have held out. It's all becoming profitable again. I should have stuck with guided transportation . . ."

GLOSSARY

A-320 An airplane built by the European consortium Airbus Industry. Designed to compete with big North American commercial aircraft, it was a major success for Charles Fiterman (q.v.).

actuator A word which, in robotics, designates any motor that exercises physical force, as opposed to an electronic circuit that conveys information.

aerotrain A mass-transit system that resembled Aramis and that underwent extensive testing from 1965 to 1976. According to the *Encyclopédie Larousse:* "The whole history of the aerotrain, which has lost government support in France, at least for the time being, after an initial period of heavy investment, is intertwined with the history of the various means of propulsion that were tried. But its history also reflects opposition from other transportation systems, railroads in particular, which saw the aerotrain as a serious rival that they did not want to exploit for their own purposes, despite proposals made along these lines." This failure, like the relative failure of the Concorde, had repercussions for the Aramis project at several points.

AIMT Automatisation Intégrale du Mouvement des Trains: A plan devised by the RATP that would completely automate the Parisian subway—that would make it, in effect, as automated as VAL. Currently, the subways are automated only in part, since each train has a driver who can take over in case of failure or emergency.

air gap A narrow break in an electromagnetic circuit. Its width affects the motor's power.

Alsthom	A large company that has long specialized in manufacturing locomotives, electrical turbines, and railway cars.
ANVAR	Agence Nationale pour la Valorisation de la Recherche ("National Agency for the Valorization of Research"). This state organization provides matching funds for innovative projects to be carried out by small private companies in collaboration with public research laboratories.
Aramis	Acronym for Agencement en Rames Automatisées de Modules Indépendants dans les Stations ("arrangement in automated trains of independent modules in stations"). The name is an allusion to one of the four heroes of Alexandre Dumas' well-known novel *The Three Musketeers.*
Araval	A transit system that, like VAL, would have been automated; but unlike Aramis, it would have run on a standard line, without a dense network, without branchings, and thus without the state-of-the-art electronics of the nominal Aramis. Orly-Val is in effect a belated realization of this "downgrading" of Aramis.
Ariane	The European space rocket, which was developed with the aim of breaking the American and Russian monopoly on satellite launchers. The enterprise is both a technological and a commercial success, and is often cited as an example of the usefulness of cooperation among European nations.
AT-2000	One of the many PRT systems dreamed up by Gérard Bardet. The direct ancestor of Aramis, it was based on the rather wild idea that train cars could be divided lengthwise into separable parts. The carrier function would thus be distinct from the feeder function, as if the aisles were detachable from the seats.
Automatisme et Technique	An engineering company, founded in 1951 by Gérard Bardet, that specialized in transport technology.
Balladur	Edouard Balladur was minister of finance in the government of Jacques Chirac (1986–1988), during the period when Aramis was being developed. He later served as prime minister, from 1993 to 1995.
Bardet	Gérard Bardet was a brilliant engineer responsible for many technological innovations, especially in the field of transportation. Trained at the Ecole Polytechnique, he was director of Automatisme et Technique (q.v.) and held the first patents on Aramis.

Bertin Jean Bertin (1919–1975), French engineer who first worked for SNECMA (q.v.), then set up his own consulting and engineering firm. With Paul Guyenne, he invented the aerotrain (q.v.).

Bienvenüe Fulgence Bienvenüe, one of the engineers who designed the Paris subway. See the excellent report produced under the direction of Maurice Daumas, *Analyse historique de l'évolution des transports en commun dans la région parisienne, 1855–1930* (Paris: Editions du CNRS, 1977).

Bus Division The "Routier": one of the two major divisions of the RATP (q.v.). The other is the "Ferroviaire," the Rail Division.

Cabinentaxi One of the many PRT systems—a cross between a train and a taxi. It was developed in Germany over a number of years, at the same time that Aramis was being built and tested.

Cantor Georg Cantor (1845–1918), German mathematician who developed set theory and demonstrated the existence of transfinite numbers.

catadiopter A type of reflector, invented by a French astronomer (at least, this is what the French claim), that is made of mirrors and lenses—i.e., that operates by means of both reflection and refraction. Catadiopters are commonly found on bicycles, cars, and other vehicles.

CET Centre d'Expérimentation Technique ("Center for Technical Experimentation," or full-system site study). This is not a fixed site but rather the name for the phase during which the chief innovations of a new technology are tested.

CGT Confédération Générale du Travail: one of France's many labor unions—the biggest in terms of membership. In tactics and ideology, it is strongly associated with the Communist Party, which was still very influential when the Aramis saga began.

Chausson A company that specializes in the manufacture of cars and buses and that made the first buses for the RATP. The Chausson APU 53 was the ancestor of the buses that run in Paris today.

Chirac Jacques Chirac, head of the Gaullist Party, served as prime minister of France 1974–1976 and 1986–1988. He was also the mayor of Paris from 1977 to 1995, and was elected president of France in 1995.

CMD Canton Mobile Déformable ("adjustable mobile

zone"). As opposed to fixed sections along the track, the CMD is a responsive zone around each car which increases or decreases depending on the car's speed and its distance from the car behind. It is the equivalent of the safe driving distance for an automobile, which varies according to speed. The CMD is the soul of Aramis (see Chapter 6).

Concorde Supersonic plane developed and built by a Franco-English consortium during the presidency of Charles de Gaulle. It was deemed a major technical success but a complete commercial failure.

Conseil Général des Ponts et Chaussées "General Council on Bridges and Roads": A supervisory board that brings together the highest officials from the prestigious three-hundred-year-old "Corps" des Ponts et Chaussées, the body that has overseen much of the French highway system since the time of Louis XIV. The Corps is made up of a small number of highly gifted students selected from the Ecole Polytechnique and sent on to specialized elite engineering schools. In the course of their careers, these chosen few go on to head various national agencies as well as the principal semiprivate industries in France. Unlike their American counterparts, the top French engineering schools such as the Polytechnique, the Ecole Nationale Supérieure des Mines, and the Ecole Nationale des Ponts et Chaussées are much more prestigious than universities.

conversational A term having nothing to do with the art of the salon. In cybernetics, it refers to a feedback loop between a command and its activator.

Dassault Marcel Dassault, hero of the French aircraft industry and founder of a vast and highly profitable business in military and civil aircraft after the First World War.

DATAR Délégation à l'Aménagement du Territoire et à l'Action Régionale, a regional planning commission. DATAR is one of the many institutions that attempt to redress the imbalance between Paris and the rest of France by planning more equitable development.

DTT Direction des Transports Terrestres ("Bureau of Ground Transportation") of the Transportation Ministry. This is the agency that oversees all related research institutes and developers.

ECA Electronique de Commande des Actuateurs ("electronic activator control"): an electronic interface between the onboard control unit (UGE) and the subsystems to be controlled in the car (doors, brakes, steering).

Eole One of the new, fully automated transit lines that are now being constructed in the middle of Paris to relieve congestion on the RER.

EPAL Etablissement Public de l'Agglomération Lilloise: a semiofficial body that oversees the creation of new townships and new infrastructure in the Lille Region in the north of France.

Espace A car designed by Matra and produced by Renault. The Espace ("Space") served as the model for the Aramis cabin.

Ficheur Michel Ficheur was the man behind VAL. Later he moved to the Research Ministry to oversee transportation research at a time when the ministry was becoming increasingly powerful. Ficheur developed a methodology for supervising research projects in a more systematic way.

Fiterman Charles Fiterman, a Communist, was France's transportation minister from 1981 to July 1984. He played a crucial role in the Aramis story.

FNAC A Paris bookstore chain that created a small-scale, short-distance transportation system for the parking garage of its store on the rue de Rennes.

Fourcade Jean-Pierre Fourcade, former minister with the UDF (a right-of-center party) and first vice-president of the Ile-de-France Regional Council.

Francilienne beltway A highway around Paris that allows through traffic to bypass the city instead of clogging its arteries.

Frybourg Michel Frybourg, a graduate of the Ecole Polytechnique, founded INRETS (q.v.) and headed it until 1982.

Giraud Michel Giraud, from the same party as Jacques Chirac, held several cabinet posts and served as president of the Ile-de-France Regional Council.

Giraudet Pierre Giraudet, the RATP's director from 1971 to 1975, was responsible for a major modernization of the metro. Under his leadership, the Paris system was largely automated—with drivers—and the last ticket puncher disappeared from the Porte des Lilas station.

guided transportation Transportation that runs on rails—subways, trolleys, commuter trains, and so on.

Habegger A minitrain developed in Switzerland in the early

1960s. Based on the monorail, with electric power and tiny cars, it was never completely automated and has been used only in amusement parks. Though it was not a fully fledged PRT, it demonstrated to transportation engineers that systems of intermediate size and complexity existed between trains and automobiles.

heterogeneous engineering Expression coined by John Law to describe the multiplicity of worlds in which an engineer must function simultaneously in order to construct an artifact. The phrase is used in opposition to the idea that engineers deal with nothing but "purely technical" matters. See John Law, "Technology and Heterogeneous Engineering: The Case of Portuguese Expansion," in W. E. Bijker, T. P. Hughes, and T. Pinch, eds., *The Social Construction of Technological Systems: New Directions in the Sociology and History of Technology* (Cambridge, Mass.: MIT Press, 1987), pp. 111–134.

homologation A technological and legal term. To say that the transportation minister and the RATP (the minister's designated contracting authority) "homologate" Aramis means that they formally declare the system to be not only feasible in principle but also safe and authorized to transport passengers.

hyperfrequencies Extremely high radioelectric frequencies, above 1,000 megahertz. Hyperfrequency makes radioelectric couplings of very high capacity possible with power of just a few watts.

Ile-de-France Region Since 1982 France has been fighting a tendency toward increasing State control and has been doing so through a process of decentralization, creating administrative "regions" with sizable powers and financial abilities. The Ile-de-France Region is one of these. But the fact that Paris lies in the middle of it means that the Region is very complicated to manage, since Paris is at once a city and the seat of national government, and is administered by both the elected mayor and a prefect designated by the government. In addition, Paris is a *département*—one of seven smaller administrative units inside the Region. Each of these administrative levels has its own responsibilities and concerns in the matter of transportation.

INRETS Institut de Recherche sur les Transports ("Institute for Transportation Research"): a research center dependent on the Transportation Ministry. INRETS examines the ministry's technical dossiers, often a difficult task given the scale of operations—both technological and financial—of the SNCF, the RATP, and the major automobile manufacturers.

isotopy — In semiotics, the set of procedures which gives the reader the impression that there is continuity among the characters and parts of a narrative. The term was imported into semiotics by A. J. Greimas; see A. J. Greimas and J. Courtès, *Semiotics and Language: An Analytical Dictionary,* trans. Larry Crist et al. (Bloomington: Indiana University Press, 1983).

Lagardère — Jean-Luc Lagardère has been president of Matra since 1977.

Lépine competition — An annual fair, conceived by a famous prefect of Paris and first held in 1901. The fair offers inventors of every stripe the opportunity to present their devices to one another and to would-be investors. Its name has become synonymous with weird and useless gadgetry.

Matra company — French high-tech company (also known as Société des Engins Matra), founded in 1945. Under the leadership of Jean-Luc Lagardère, it has become preeminent in its field, with major divisions devoted to military applications, telecommunications, automobile technology, and mass transit. It is also involved in many activities relating to publishing and the media. Its branch Matra Transport is one of the central characters in this story, since it was set up to handle the business generated by VAL and, at least in part, by Aramis.

Matra Transport — A subsidiary of the Matra company founded in 1972 to develop the VAL system in Lille, as well as Aramis. It has since become an important company in its own right, selling automated-transport systems around the world (e.g., the city of Taipei, and Chicago's O'Hare Airport).

Mauroy — Pierre Mauroy was prime minister from 1981 to 1984. He was replaced by Laurent Fabius up to the 1986 legislative elections, which were won by the Right.

Mirage III — One of the most sophisticated of the Dassault fighter planes.

Mitterand — François Mitterand, a moderate leftist, was elected president of France in 1981. He won reelection for a second seven-year term in 1988.

module A unit that is considered to be reusable in other transportation systems without any specific adaptation to the particular circumstances of a given site. "Modular" research is the opposite of "specific" research. In practice, the complexities of guided transportation make a modular approach very difficult.

MST Mission Scientifique et Technique ("Scientific and Technical Mission.") Branch of the Research Ministry responsible for the politics of science.

Notebart In the 1970s and 1980s Arthur Notebart was president of the Lille urban community, mayor of the city, and a member of the Chamber of Deputies. He put all his energy into the VAL project.

orange card A special fare card that allows Parisians to pay just once a month, or even once a year, for unlimited use of public transportation within the city and its immediate outskirts. Half the cost of the card is borne by the passenger's employer. This rate structure makes it difficult to calculate the profitability of a new investment—but it does make public transportation in Paris relatively inexpensive, indeed a real bargain compared to other costs (as tourists can hardly fail to note). The low fees are the result of a decision by Pierre Giraudet (q.v.) to simplify the subway- and train-fare system.

Orly Rail Traditional rail system that links Paris with Orly Airport. It has one major disadvantage: passengers have to take a bus to reach the terminals.

Orly-Val A VAL system a few kilometers long that was opened in 1990 to link Orly Airport to the RER station in the Paris suburb of Antony. Built with private funding, it quickly went bankrupt—as predicted—and is now run jointly by the RATP and the SNCF.

PCC Poste Central de Commande ("central command post"): supervisory hub of the Aramis system. It includes a computer network, workstations, consoles, video monitors, and telephones that permit electronic control and surveillance of the system's components, stations, and movements, as well as communication with maintenance, police, and emergency services.

Petite Ceinture An old rail line that circles Paris and that, for much of its length, has fallen into disrepair. See map in the frontmatter.

phonic wheel A cogged wheel that allows very fine calibrations of its axle's rotation.

Poma 2000 — A PRT system which is based on the same principle as the ski lifts at mountain resorts and the cable cars in San Francisco. Each car, which is small and has no driver, is automatically linked to a continuously running cable and circulates between stations. The system has been operating for several years in the French town of Laon.

Prometheus — The code name for a European Community project to develop "smart cars."

PRT — Personal Rapid Transit: the adaptation of mass transportation to the needs of the individual or small groups of passengers, in order to fight the hegemony of automobile transportation. The concept of PRT became very popular during the Kennedy administration, but it has since fallen out of favor somewhat. Aramis was the longest-running PRT project. In recent years there has been a renewal of interest in this technology—a hybrid of car and train.

Quin — Claude Quin was a member of the Communist Party, an elected representative from Paris. He headed the RATP from 1981 to 1986.

Rafale — A military aircraft developed by Dassault amid much controversy over its cost and specifications. Like Aramis, it was a symbol of France's high-tech ability.

Rail Division — The "Ferroviaire": one of the two major divisions of the RATP (q.v.). The other is the "Routier," the Bus Division.

RATP — Régie Autonome des Transports Parisiens: the operating agency responsible for the subways and buses of the city of Paris and its immediate outskirts. Its status is intermediate between that of a private enterprise and that of an administration. Like the SNCF (q.v.), it functions as both buyer and client: it is responsible not just for transporting passengers but for designing, commissioning, and ordering the means of transportation. The ministry that supervises it has only limited control over its operations. The agency is the product of a merger between two companies—one involved with aboveground transport and the other with underground transport—which had completely different missions and technical cultures. This difference is still apparent in the fact that the RATP has two branches—the Rail Division ("Ferroviaire") and the Bus Division ("Routier")—and it played a role in the Aramis case, since Aramis is a hybrid creature, both *ferroviaire* (it runs on tracks) and *routier* (it has a buslike construction). See Michel Margairaz. *Histoire de la RATP: La singulière aventure des transports parisiens* (Paris: Albin Michel, 1989).

refresh time — The speed with which the data in a computer or on a screen are brought up to date.

RER — Réseau Express Régional ("Regional Express Network"), a mass-transit system that is saturated during rush hours in its central Paris stations, especially on Line A, which runs east-west. Two major competing projects are slated to reduce the pressure on the RER network: the RATP's Meteor (an automated metro system for which Matra won the contract), and the SNCF's Eole.

Roissy Rail — A traditional rail system that links Paris with Charles de Gaulle Airport. Its major disadvantage is that passengers must take a bus to reach the terminals.

SACEM — An automated guidance system used on traditional metros (RER). It allows trains to approach each other more closely than human drivers can permit them to do: it overrides the safety instructions and enables trains to follow each other at the hair-raising interval of one and a half minutes. It was developed at the same time as Aramis and involves much the same kind of software.

Saunier-Seïté — Alice Louise Saunier-Seïté, right-wing minister of higher education and research; the *bête noire* of researchers, who compared her to Margaret Thatcher.

Schneider — A manufacturing company that made some of the RATP's first buses.

SK — A short-distance transportation system, a kind of horizontal cable car much like the one in Morgantown, West Virginia. The SK can be seen in the Villepinte Exhibition Park north of Paris.

SNCF — Société Nationale des Chemins de Fer: the French state-owned railway company. Like the RATP, it is responsible for transporting passengers, as well as for designing means of transportation, having them constructed, and ordering them. It thus combines the functions of buyer and client, and enjoys a large measure of administrative autonomy.

SNECMA — Société Nationale d'Etude et de Construction de Moteurs d'Aviation ("National Company for the Study and Construction of Aircraft Engines"): a major high-tech French company specializing in airplane and rocket engines.

sniffer plane An airplane that was the focus of a huge scandal in the early 1980s. Oil companies and bogus scientists collaborated in financing and developing a revolutionary system designed to sniff out oil reserves from above. The instrument's black box, when it was finally opened, turned out to be empty. The scandal was publicly acknowledged in 1983, after being denied several times by former president Valéry Giscard-d'Estaing.

SOFRES One of France's largest polling firms, specializing in political and marketing polls.

SOFRETU A consulting and development firm for guided-transportation systems. The RATP owns 80 percent of it.

South Line The Rocade Sud: a major transverse line into the suburbs of Paris (see map in the frontmatter) which allowed people to travel, without passing through Paris, among many small communities where demand was too small to support a standard metro system.

testing phase The last phase of a project before final approval—that is, acceptance by the public authorities of the transportation system's safety.

TGV Train à Grande Vitesse ("high-speed train"). These trains, which criss-cross France at a speed of 300 kilometers per hour, have been a great technical and economic success for the SNCF.

TRACS A very clever system of moving sidewalks that succeed in the apparently impossible feat of going faster in the central portion than at either edge. It was abandoned in 1986. Since the same engineers and the same decisionmakers work on all such transportation systems, successes or failures in one area have repercussions for the others, even though the technological principles may be quite different.

UGE Unité de Gestion Embarquée ("onboard control unit"): an electronic unit on board the transportation vehicle that functions to control safety, driving, and supervision.

UGT Unité de Gestion de Tronçon ("traffic-control unit"): one of Aramis' lower-tier electronic control systems, in charge of dispatching the various vehicles. For each one-kilometer sector of track, the UGT controls the anticollision function, exchanges data with the Central Command Post, regulates merging of the cars, monitors the cars' positions, and transmits alarm signals when necessary.

VAL An automated subway similar to the shuttle that serves the Atlanta airport. First implemented in Lille, a large city in northern France, it has since been exported to several countries, including the United States (it is now being used at Chicago's O'Hare Airport).

valorimeter Any instrument that measures a value—for example, in units of money, in kilocalories, or in petroleum-equivalents. It designates any measuring instrument that *establishes an equivalence* between dissimilar things.

variable-reluctance motor The main mechanical invention of Aramis. The rotary engine allows engineers to do away with the gears that, in ordinary vehicles, link the rotation of the electric motor with the axles. By varying the reluctance—which is, in magnetism, what resistance is in electricity—the engineers connect the functions of axle and rotor.

Villette (La) A slaughterhouse in the north of Paris. It was a failure, in technical as well as economic terms—a real pork barrel. The building was converted into a gigantic Museum of Science and Industry, four times the size of the Beaubourg museum. It is now a huge monument devoted to the "glory of Science and France."